MATERIALS SCIENCE AND TECHNOLOGIES

POLYCRYSTALLINE MATERIALS

SYNTHESIS, PERFORMANCE AND APPLICATIONS

MATERIALS SCIENCE AND TECHNOLOGIES

Additional books in this series can be found on Nova's website under the Series tab.

Additional e-books in this series can be found on Nova's website under the eBooks tab.

MATERIALS SCIENCE AND TECHNOLOGIES

POLYCRYSTALLINE MATERIALS

SYNTHESIS, PERFORMANCE AND APPLICATIONS

JIM OLSON
EDITOR

nova
science publishers
New York

Library of Congress Cataloging-in-Publication Data

ISBN: 978-1-53613-864-1

Published by Nova Science Publishers, Inc. † New York

CONTENTS

PREFACE

Polycrystalline materials, both dense and porous, are traditionally fabricated from ceramic, metal or composite powders via powder technology routes that include shaping and sintering steps. During shaping a compact body is formed from the granular raw material, while sintering leads to strengthening of this compact and coarsening. In this collection, the authors summarize the basic concepts and classical models used for describing solid state sintering processes. Following a brief introduction, in the second section a historical overview of early sintering theory is given, followed in the third section by an explanation of the driving force of solid state sintering. Following this, grain boundary is analyzed to discern how the spatial orientation of grains themselves may be defined generally and exactly. Additionally, one can easily understand that there is a correlation between the change in the grain orientation and grain deformation and this correlation was mathematically demonstrated. Finally, the authors interpret the quantity of grain orientation corresponding with the experimental measured value, the analysis of local deformation could be performed experimentally on the basis of their results. In closing, following an outline describing the historical development of ECAS and related techniques, the principle of ECAS is briefly described, current opinions concerning sintering mechanisms are summarized, and the problem of carbon contamination is discussed in some detail.

Chapter 1 - Polycrystalline materials, both dense and porous, are traditionally fabricated from ceramic, metal or composite powders via powder technology routes that include shaping and sintering steps. During shaping a compact body is formed from the granular raw material, while sintering leads to strengthening of this compact (via neck growth and grain shape changes) and usually also to densification (porosity and pore size reduction) and coarsening (grain growth). Generally both the raw materials and the processing procedures determine the final microstructures and properties of the resulting polycrystalline materials. The paradigmatic model case, often considered as a benchmark, is solid state sintering (SSS), i.e., sintering without the transient assistance of a liquid melt phase that usually remains in the microstructure as a glassy (amorphous) or a second crystalline phase after cooling. In fact, SSS is the via regia to pure single-phase polycrystalline materials. In this contribution the authors summarize the basic concepts and classical models used for describing SSS processes. Following a brief introduction, in the second section a historical overview of early sintering theory is given, followed in the third section by an explanation of the driving force of SSS. The fundamental significance of the dihedral angle in SSS is explained, as well as the relation between surface curvature and the Kelvin-type equations for vapor pressure and vacancy concentration. In the fourth section sintering stages and mechanisms are discussed. Densifying and non-densifying sintering mechanisms are explained, with a special focus on diffusion mechanisms (surface, grain boundary and volume diffusion) and their temperature dependence (described via Arrhenius equations). The different stages of sintering are treated using the two-sphere model for neck growth in the initial stage and the tetrakaidecahedral cell model (Kelvin cell model) for densification, shrinkage and pore shape changes in the intermediate and final stages. Selected aspects of sintering kinetics and its monitoring by shrinkage and specific surface measurements are briefly outlined as well. In the fifth section grain growth models are briefly summarized and current topics of research in this field are mentioned. In the context of the evolution of grain size distributions during SSS the importance of the Rayleigh distribution is indicated and the necessity to account for the random section problem in order to obtain three-dimensional grain size distributions from two-dimensional sections is emphasized. Finally the sixth section gives a critical account of the concept of master sintering curves, which has turned out to be a successful tool for sintering practice.

Chapter 2 - Grain boundary - i.e., the interface between individual grains is one of the principal microstructural parameters in the polycrystalline materials. In case of isotropic structure, the grains have isometric dimensions and mean grain size or grain size distribution and specific surface area of grain boundaries are sufficient structural characteristics. In case of anisotropic structure the grains have anisometric dimensions caused, for instance, by plastic deformation and it is necessary to describe their orientation. In this context the first question is how can the spatial orientation of grains themselves be defined generally and exactly, the second question is how can the strain be estimated from this orientation. It is possible to suspect intuitively that grain surface orientation in this sense can be evaluated by the prevailing orientation of grain surface area elements. The anisotropic microstructure can be decomposed into isotropic, planar and/or linear oriented components - specific surface area of grain boundaries and these parameters are experimentally measured using stereology. But real state of grain shape is quite impossible to describe, therefore model of grain shape based on an idealized shape (tetrakaidecahedron) of grains has been proposed. One can easily understand that there is a correlation between the change in the grain orientation and grain deformation and this correlation was mathematically demonstrated. Finally, the authors interpret the quantity of grain orientation corresponding with the experimental measured value, the analysis of local deformation could be performed experimentally on the basis of their results.

Chapter 3 - Electric current assisted sintering (ECAS), also called pulsed electric current sintering (PECS) or field assisted sintering technique (FAST), but more widely known under the popular – albeit less appropriate – name spark plasma sintering (SPS), has gained considerable attention from the scientific community during the last few decades, because it is a convenient method for the fast densification of otherwise hard-to-sinter ceramic powders, both oxide and non-oxide. While in the case of metal powders the densification is always affected by the flow of electric current through the sample itself (initially via the particle contact points and subsequently via the sinter necks), this is usually not the case for ceramic powders, at least not for temperatures below which ionic conductivity plays a significant role. Therefore, in contrast to metals, in ceramics it is essentially only the Joule heating of the mold (usually a graphite mold) and the ensuing high heating rate that distinguishes ECAS from other forms of (uniaxial) hot pressing. Thus, also the sintering mechanisms in ECAS are not too different from those of (uniaxial) hot pressing. Since ECAS is an extremely versatile method, allowing the simultaneous control of many process parameters and conditions, such as electric current levels and pulse rates, temperature schedules (including heating and cooling rates, dwell times and hold temperatures), pressure schedules (including pressure increase and decrease rates, dwell times and hold pressures) and atmospheres (usually vacuum or inert), the number of actual and potential applications of ECAS is steadily increasing and already now too large to be treated in a single review. Therefore, the present contribution focusses on recent ECAS achievements and progress in the special field of translucent and transparent ceramics. For the preparation of these materials appropriate grain size, extremely high density and the minimization (preferably absence) of second phase inclusions are critical factors. Following an outline describing the historical development of ECAS and related techniques, the principle of ECAS is briefly described, current opinions concerning sintering mechanisms are summarized, and the problem of carbon contamination is discussed in some detail. In this context the use of boron nitride as a protective coating and lithium fluoride as a sintering aid is mentioned. Fundamental optical properties are defined, an overview of microstructural factors affecting transmittance is given, and literature data (ECAS conditions and resulting optical properties) are compared for various oxide (and a few silicate, phosphate or non-oxide) ceramics that have so far been prepared in transparent or at least translucent form for wavelengths close to the visible or infrared range of the electromagnetic spectrum. The materials reviewed in this chapter include, among others, alumina, lutecia, magnesia, yttria, zirconia, titanates, zirconates, lutetium compounds, magnesium aluminate spinel (magnesia alumina spinel), yttrium aluminum garnet (YAG), hydroxyapatite, mullite, aluminum nitride, aluminum oxynitride, silicon aluminum oxynitride, calcium fluoride, magnesium fluoride, sphalerite, and yttria-magnesia nanocomposites.

In: Polycrystalline Materials
Editor: Jim Olson

ISBN: 978-1-53613-864-1

Chapter 1

Basic Concepts and Classical Models of Solid State Sintering

Willi Pabst*, Tereza Uhlířová, Vojtěch Nečina and Eva Gregorová
Department of Glass and Ceramics,
University of Chemistry and Technology,
Prague (UCT Prague), Prague, Czech Republic

Abstract

Polycrystalline materials, both dense and porous, are traditionally fabricated from ceramic, metal or composite powders via powder technology routes that include shaping and sintering steps. During shaping a compact body is formed from the granular raw material, while sintering leads to strengthening of this compact (via neck growth and grain shape changes) and usually also to densification (porosity and pore size reduction) and coarsening (grain growth). Generally both the raw materials and the processing procedures determine the final

---

* Corresponding Author: Prof. Dr. Willi Pabst, Department of Glass and Ceramics, University of Chemistry and Technology, Prague (UCT Prague), Techniká 5, 166 28 Prague 6, Czech Republic. Email: pabstw@vscht.cz.

microstructures and properties of the resulting polycrystalline materials. The paradigmatic model case, often considered as a benchmark, is solid state sintering (SSS), i.e., sintering without the transient assistance of a liquid melt phase that usually remains in the microstructure as a glassy (amorphous) or a second crystalline phase after cooling. In fact, SSS is the via regia to pure single-phase polycrystalline materials. In this contribution we summarize the basic concepts and classical models used for describing SSS processes. Following a brief introduction, in the second section a historical overview of early sintering theory is given, followed in the third section by an explanation of the driving force of SSS. The fundamental significance of the dihedral angle in SSS is explained, as well as the relation between surface curvature and the Kelvin-type equations for vapor pressure and vacancy concentration. In the fourth section sintering stages and mechanisms are discussed. Densifying and non-densifying sintering mechanisms are explained, with a special focus on diffusion mechanisms (surface, grain boundary and volume diffusion) and their temperature dependence (described via Arrhenius equations). The different stages of sintering are treated using the two-sphere model for neck growth in the initial stage and the tetrakaidecahedral cell model (Kelvin cell model) for densification, shrinkage and pore shape changes in the intermediate and final stages. Selected aspects of sintering kinetics and its monitoring by shrinkage and specific surface measurements are briefly outlined as well. In the fifth section grain growth models are briefly summarized and current topics of research in this field are mentioned. In the context of the evolution of grain size distributions during SSS the importance of the Rayleigh distribution is indicated and the necessity to account for the random section problem in order to obtain three-dimensional grain size distributions from two-dimensional sections is emphasized. Finally the sixth section gives a critical account of the concept of master sintering curves, which has turned out to be a successful tool for sintering practice.

Keywords: polycrystalline materials, solid state sintering (SSS), densification, grain growth, sintering stages, sintering kinetics, sintering mechanisms, surface diffusion, grain boundary diffusion, volume (lattice diffusion), two-sphere model, Coble model, spherical particle shape, tetrakaidecahedral grain shape, Kelvin cells, relative density, porosity, pore size, grain size distribution, Hillert distribution, Rayleigh distribution, master sintering curve (MSC)

1. INTRODUCTION

Sintering is a high-temperature and possibly high-pressure process that transforms particulate or granular matter into more or less strong and more or less dense polycrystalline solid compacts upon heating. In the widest sense it is a process that can be classified as a branch of powder technology. However, there is no universally accepted definition of sintering. Depending on the individual point of view and on the definition adopted one may argue that sintering has been used by mankind since the beginning of ceramic technology, i.e., for thousands of years. However, in most cases the manufacture of traditional silicate ceramics involves a considerable amount of glass melt that is liquid at high temperature and solidifies into a glass phase upon cooling, unless the cooling rate is slow enough and the composition suitable for recrystallization to occur. Although sintering in the presence of a liquid phase (called liquid phase sintering or, when the content of glass melt is high, vitrification) occurs not only in the context of traditional silicate ceramics but also in many advanced ceramics, and although some solid state sintering mechanisms may be active in the presence of a liquid phase (glass melt) as well, they are usually overshadowed by capillary effects and complex interactions, or even chemical reactions, between solid and liquid phases. Therefore sintering in the presence of a liquid phase tends to conceal the most interesting key feature of sintering which consists in the fact that strengthening, densification and coarsening can occur without the presence of a liquid phase. Actually solid state sintering (SSS) is the type of sintering process that exhibits this feature in its clearest form.

It is the purpose of this chapter to present the basic concepts and classical models of SSS. However, this chapter will not simply repeat information that can be found elsewhere, for example in the authoritative treatments by German (German 1996) or Rahaman (Rahaman 2003) or the more recent multiauthor book edited by Fang (Fang 2010), but it is a critical and updated account of the current state of the art in SSS theory, largely based on a study of both classical and recent papers cited in the list of references. A glance on the research activities in the field of sintering (especially SSS) since 1945, see Table 1, shows that the interests have been very much focused on grain growth kinetics, compared to which the research in densification kinetics, for example, has been less intense. Although this finding cannot be discussed away, statistics of this type have to be read with great caution, because they can easily be used to distort the real situation and the actual needs. It can be seen, for example, that there are much more papers and citations concerning SSS mechanisms than concerning SSS kinetics and stages, although the latter two are much more important from the practical point of view. A tentative explanation for this is that hypotheses on SSS mechanisms are readily invoked in different contexts, sometimes without the necessity to prove them. *Cum grano salis* it might be said that postulating mechanisms is easier than measuring kinetics. Similarly, it might be surmised that grain growth kinetics is just an easier topic than densification kinetics, both from the viewpoint of theory and from the viewpoint of computer simulation. Indeed it will be shown below that, compared to grain growth, densification is a highly complex topic that does not lend itself readily to unification. Concerning the concept of master sintering curves, a naive interpretation of the scientometric data would attribute to them not more than peripheral importance, although in reality they are one of the hot topics of current research in sintering today. The reason is that the concept of master sintering curves is a relatively recent concept, available for not much more than two decades now, while the other topics are much more traditional and much broader. Indeed, closer inspection reveals that the concept of master sintering curves constitutes a quite significant part of research on SSS kinetics today. All the topics listed in Table 1 will be treated in this chapter.

In the second section a historical overview of early sintering theory is given, followed in the third section by an explanation of the driving force of SSS in terms of Gibbs free energy reduction, dihedral angle and interface curvature. The fundamental significance of the dihedral angle in SSS is explained, as well as the relation between surface curvature and the Kelvin-type equations for vapor pressure and vacancy concentration. In the fourth section sintering stages and mechanisms are discussed. Densifying and non-densifying sintering mechanisms are explained, with a special focus on diffusion mechanisms (surface, grain boundary and volume diffusion) and their temperature dependence (described via Arrhenius equations). The different stages of sintering are treated using the two-sphere model for neck growth in the initial stage and the tetrakaidecahedral cell model (Kelvin cell model) for densification, shrinkage and pore shape changes in the intermediate and final stages. Selected aspects of sintering kinetics and its monitoring by shrinkage and specific surface measurements are briefly outlined as well. In the fifth section grain growth models are summarized and current topics of modeling are mentioned. In the context of the evolution of grain size distributions during SSS the importance of the Rayleigh distribution is indicated and the necessity to account for the random section problem in order to obtain three-dimensional grain size distributions from two-dimensional sections is emphasized. Finally the sixth section gives a critical account of the concept of master sintering curves, which has turned out to be a successful tool for sintering practice.

Table 1. Scientometric data for different topics related to solid state sintering (SSS) since 1945 (after Web of Science Core Collection, Thomson-Reuters, accessed 3 March 2018)

Topic	Total number of publications	Average citations per paper	H-index (Hirsch index)	Total number of citations
SSS mechanisms	1104	16.36	52	18056
SSS kinetics	379	20.92	43	7928
SSS stages	323	16.02	36	5175
Sintering kinetics	4661	16.12	98	75141
Densification kinetics	1450	17.59	64	25503
Grain growth kinetics	7316	18.15	123	132765
Master sintering curves	158	11.72	22	1851

2. Historical Development of Sintering Theories

The historical origins of solid state sintering (SSS) have to be sought probably in the early powder metallurgy of platinum (Chaston 1980). However, only from the second half of the 20th century onwards sintering theories have been developed that could be applied to ceramic technology. Frenkel's seminal paper on the viscous flow of crystalline bodies under the action of surface tension (Frenkel 1945) and Zener's key finding, reported by Smith (Smith 1948), that grain growth in polycrystalline metals is impeded by immobile second-phase inclusions, resulting in a relation of the type

$$\phi_{inclusion} \propto \frac{D_{inclusion}}{G}, \tag{1}$$

where $\phi_{inclusion}$ is the volume fraction of the inclusions, G the grain size and $D_{inclusion}$ the size of inclusions, are among the first important achievements in scientific sintering theory. It should be mentioned, however, that Frenkel's paper concerns the sintering mechanism of viscous flow, a mechanism that is commonly encountered in the sintering of glass particles (or so-called glass frit), but not in solid state sintering (SSS) of crystalline materials. In the case that the obstacles (inclusions) are exclusively mobile pores at grain boundaries, Zener's relation has to be replaced by the relation (Burke & Rosolowski 1975)

$$\phi \propto \left(\frac{P}{G}\right)^2, \tag{2}$$

where ϕ is the intergranular porosity (i.e., the volume fraction of pores), P the intergranular pore size and G the grain size as above. It is clear that in the case of non-spherical pores and grains, and/or a size distribution of either, the size measure has to be appropriately defined, for example via an equivalent diameter or a statistical length measure (e.g., a mean chord length). Therefore it makes no sense to specify the constants of proportionality in these relations. It is evident, however, that in polycrystalline materials the intergranular porosity, intergranular pore size and grain size is correlated in some way[1] and that the ratio between pore size and grain size is in some way related to the final density achieved during sintering.

Another early milestone of sintering theory was the paper on the phenomenological theory of sintering by Mackenzie and Shuttleworth (Mackenzie & Shuttleworth 1949). The results of this paper, although not explicitly restricted to amorphous solids[2] and successfully applied also to final stage sintering of polycrystalline solids (Clark et al. 1953), see also (Hlaváč 1988), have been immediately applied to the sintering of powdered glass (Clark & White 1950) and are today applied mainly for describing the final-stage sintering of amorphous powders (Rahaman 2003), while the Frenkel model is the model of choice for initial-stage sintering of amorphous powders (Hlaváč 1988, Rahaman 2003). A corresponding intermediate-stage sintering model for amorphous powders has been developed much later by Scherer (Scherer 1977a, 1977b, 1984). Actually these three models for describing the different sintering stages of amorphous powders exhibit considerable overlap and lead to similar predictions of the relative density as a function of the (appropriately normalized) sintering time. This is taken to be an indication of the relative insensitivity of the sintering behavior of amorphous materials to microstructural details (Rahaman 2003). It is clear that the practical significance of the Frenkel, Scherer and Mackenzie-Shuttleworth models lies in the field of sintering processes dominated by glass melts. Apart from the sintering of powdered glass (glass frits), where the sintered

---

[1] The intergranularity of the pores has to be emphasized here, because such a correlation does not exist for pores significantly larger than the grains, e.g., in porous ceramics made with pore-forming agents or in highly porous cellular materials, such as ceramic foams.

[2] It should be noted that also Frenkel's work (Frenkel 1945) was not explicitly limited to amorphous solids, and was actually intended to be a sintering model for crystalline solids; such an interpretation of Frenkel's results, albeit not universally accepted, is indeed thinkable, because on the atomistic scale the viscosity coefficient is related to the diffusion coefficient (Kuczynski 1949).

product does not contain grain boundaries at all, these are mainly found in the field of liquid phase sintering (LPS) with high contents of glass melts, i.e., processes that are sometimes called vitrification (Kingery et al. 1976, Rahaman 2003). Processes of this type are beyond the scope of this chapter on solid state sintering (SSS).

It seems that the idea to admit diffusion as a sintering mechanism has for the first time been suggested by Pines (Pines 1946), see also (Geguzin 1981). However, the first quantitative work to take into account diffusion as a possible mechanism and to use the diffusion equation for calculating sintering kinetics was Kuczynski's paper on self-diffusion in the sintering of metallic particles (Kuczynski 1949). In this paper Kuczynski applied a simple model of a spherical particle in contact with a planar surface of the same material to calculate the initial-stage kinetics of isothermal neck growth, i.e., the increase of the neck diameter X with time t (at constant temperature), not only for viscous or plastic flow, but also for evaporation and condensation (based on the Kelvin equation for the vapor pressure of the solid) and for volume and surface diffusion (based on a Kelvin-type equation for the vacancy concentration). His main results can be summarized in the form (Hlaváč 1988)

$$\frac{X^n}{G^\mu} \propto t \,, \qquad (3)$$

where the neck growth exponent n is 2, 3, 5 and 7 for viscous or plastic flow, evaporation and condensation, volume diffusion, and surface diffusion, respectively, and the corresponding grain size exponents μ are 1, 1, 2 and 3, respectively. Apart from these key results, Kuczynski's paper (Kuczynski 1949) contains at least three other important findings: first, the recognition that even in metals like copper the sintering process is governed by diffusion and not by plastic deformation (or viscous flow), second, a convincing argumentation in favor of vacancy diffusion, as opposed to other diffusion mechanisms (such as diffusion of interstitial atoms or atom exchange) and, third, the experimental finding that surface diffusion is operative in the early stage of sintering, whereas volume diffusion is more important in the later stage or for higher temperature or for larger grains. At that time it was already evident that the grain size (and thus also the initial particle size) plays a crucial role in sintering and determines the sintering rates. One year later, Herring published his famous paper (Herring 1950a) in which he presented a scaling law relating grain size ratios to the inverse ratios of sintering rates (or the corresponding ratios of sintering times Δt to achieve certain relative densities). Herring's scaling relation can be written as

$$\frac{\Delta t_1}{\Delta t_2} = \left(\frac{G_1}{G_2}\right)^M, \tag{4}$$

where the scaling exponent M is 1, 2, 3 and 4 for viscous or plastic flow, evaporation- condensation, volume diffusion and surface diffusion, respectively. The basic physical content of this scaling relation is clear: finer powders are expected to sinter more readily. Dealing with non-dimensional ratios, Herring's scaling law does not assume a specific geometrical model. It is therefore principally applicable to any particle shape and all sintering stages (Rahaman 2003). Its derivation, however, is based on the assumption that the two systems to be compared are geometrically similar initially and remain similar during sintering, so that the scaling is just a change in the absolute dimensions of the system. Scaling laws of this type seem to be more successful for the sintering of metallic than of ceramic powder compacts (Rahaman 2003).

Based on Kuczynski's work (who compared theoretical predictions with experimental results for copper and silver) and Herring's scaling laws, Kingery and Berg published a key paper in which they presented calculations of neck growth kinetics for initial-stage sintering on the basis of a two-sphere model (Kingery & Berg 1955). Comparing theoretical predictions with experimental results they found that the sintering mechanisms of glass spheres, sodium chloride spheres and copper spheres were viscous flow, evaporation-condensation and diffusion, respectively, while for alumina and stabilized zirconia the results remained inconclusive. Important contributions of this paper were the clear distinction between densifying and non-densifying sintering mechanisms and – for the former – the newly suggested connection between volumetric or linear shrinkage and sintering time. The latter enabled the metallurgist or ceramicist to replace difficult microscopic measurements of sinter necks between individual particles by simple macroscopic measurements of body shrinkage. This practical approach is based on the approximation (Burke 1958, Kingery & Berg 1955)

$$\frac{\Delta L}{L_0} = \frac{h}{G} \approx \left(\frac{X}{2G}\right)^2, \tag{5}$$

where h is half the interpenetration distance of the spheres, see also (Exner 1979, Exner & Petzow 1975, German 1996, Hlaváč 1988, Rahaman 2003, Uskoković & Exner 1977). It has to be noticed that Kingery and Berg (Kingery & Berg 1955) claim validity of these relations only for small shrinkage (expecting reasonable validity for volume shrinkage up to 2% and complete failure for volume shrinkage higher than 6%), and for such small shrinkage values their approximation

$$\frac{\Delta V}{V_0} \approx 3 \cdot \frac{\Delta L}{L_0} \tag{6}$$

for transforming volume shrinkage into linear shrinkage is indeed justified. However, it seems to be a common misunderstanding in the current literature to assume that this approximation is automatically valid if the microstructure is isotropic. Clearly, isotropy is a necessary condition for Equation 6 to be valid, but not a sufficient one. The exact relation is (German 1996)

$$\frac{\Delta V}{V_0} = 1 - \left(1 - \frac{\Delta L}{L_0}\right)^3 . \tag{7}$$

Another achievement of the Kingery and Berg paper (Kingery & Berg 1955), compared to Kuczynski's paper (Kuczynski 1949), was the recognition that the grain boundaries, and not the free surfaces, are the sinks of vacancies. However, the hope that measurements of sintering kinetics could be used to determine diffusion coefficients was fulfilled only partially (for metals rather than for ceramic oxides), see also (Hlaváč 1988).

Based on Kuczynski's 1949 paper (Kuczynski 1949) and the 1955 paper by Kingery and Berg (Kingery & Berg 1955), which both dealt exclusively with the initial stage of SSS (where the neck-to-grain-size ratio is below 0.3, i.e., $X/G < 0.3$), Burke (Burke 1957, 1958) focused on the role of grain boundaries in sintering (Burke 1957), summarized the state of the art at that time and focused attention to grain growth, both normal (continuous) and abnormal (discontinuous), during the later stages of sintering (Burke 1958).

While early work on sintering theory grew out of powder metallurgical tradition, with plastic deformation and vacancy annihilation at dislocations being widely discussed topics, from the second half of the 1950s onwards (essentially beginning with Kingery's activity at the Massachusetts Institute of Technology/MIT, Cambridge, Massachusetts) sintering became a major topic in ceramic science and engineering. In particular, the interest in the sintering of ceramics obtained a fresh impetus from the late 1950s attempts (at General Electric/GE, Schenectady, New York) to develop translucent alumina ceramics (later called Lucalox®) for high-pressure sodium vapor lamps (Burke 1990), which were to become the most popular street lighting all over the world.

The *annum mirabile* of sintering theory was probably 1961. In that year, Coble published his two key papers on the sintering of crystalline solids (Coble 1961a, 1961b), in which he clearly distinguished initial, intermediate and final stages of sintering and introduced the tetrakaidecahedral grain shape model for calculating sintering kinetics in these stages. It seems that Coble was the first to fully recognize the importance of grain boundary diffusion in addition to volume and surface diffusion[3] and to realize the complex interplay between densification and grain growth during sintering. I In the same year Lifshitz and Slyozov (Lifshitz & Slyozov 1961) as well as Wagner (Wagner 1961) published (independently of each other) their papers on particle growth. Both papers emphasized the importance of size distributions, and the approach proposed therein was to become, under the name Lifshitz-Slyozov-Wagner model (LSW model), a fundamental benchmark in modern particle growth theory and at the same time the basis of later grain growth models. A milestone of similar importance was Hillert's paper on grain growth (Hillert 1965), in which he applied a LSW-type model approach for the first time to the growth of polyhedral grains in dense single-phase polycrystalline materials.[4] More recently, theories of normal grain growth in pure single-phase systems have been summarized by Atkinson (Atkinson 1988). It should be mentioned that early sintering theories and grain growth models (Burke & Turnbull 1952, Burke 1958) as well as computer simulations have tried to explain three-dimensional grain growth based on two-dimensional models. The basis of this approach is the fact that grain growth occurs by the migration of grain boundaries. The key assumption is that the latter is driven by the surface curvature of the grains (resulting in so-called curvature-driven grain growth kinetics) and is directed always toward their centers of curvature. Of course, a two-dimensional approach is simpler than a three-dimensional approach from the viewpoints of understanding and computation. In particular, an early theoretical result, the so-called von-Neumann-Mullins law (von Neumann

---

[3] A few years later, Johnson (Johnson 1969a, 1969b) was one of the first to apply computer simulations to sintering and the first who succeeded in calculating the diffusion coefficients for all three diffusion mechanisms, *viz.* volume, surface and grain boundary diffusion. He verified his results by comparing them with radioactive tracer self-diffusion data for silver. However, for oxides the approach has turned out to be less successful (Hlaváč 1988, Rahaman 2003).

[4] It has to be noted that the original LSW model has been derived for the growth of spherical particles dispersed in a liquid medium or spherical inclusions in a matrix assuming infinite dilution (non-interaction approximation).

1952, Mullins 1956), discovered in the context of two-dimensional bubble growth in soap froth or foam (Smith 1952), has been an important guide for later grain growth theories (Hillert 1965, Louat 1974, Atkinson 1988). This law, which was of fundamental importance for Hillert's application of the LSW model to polycrystalline materials, states that the two-dimensional growth rate of polygons (more precisely the polygon area, corresponding to the section area of grains) A is proportional to the number of polygon sides s minus 6,

$$\frac{dA}{dt} \propto s - 6. \tag{8}$$

This implies that the growth of two-dimensional polygons comes to an end as soon as the number of sides becomes six, i.e., in two dimensions hexagons are stable polygon shapes. This condition is a consequence of the interplay between topological constraints (area-filling polygons in two dimensions) and surface tension equilibrium (or edge tension equilibrium in two dimensions). The von-Neumann-Mullins law implies that the area of larger polygons, i.e., polygons with a larger number of neighbors, grows at a faster rate (compared to smaller polygons), and that the growth rate is fastest for the largest grains, whereas the area of small polygons decreases the faster the smaller they are. In qualitative terms some of these conclusions can immediately be transferred to three-dimensional space-filling grains in polycrystalline materials, but drawing quantitative conclusions is very problematic. For more recent attempts to find a three-dimensional analogue of the von-Neumann-Mullins law see (Glicksman 2005, Hilgenfeldt et al. 2001, Mullins 1989). An important result of this research is that polyhedra with less than 14 faces tend to shrink and those with more than 14 faces tend to grow (Glicksman 2005). It is generally acknowledged today that grain growth processes (coarsening) lead after sufficiently long time (asymptotically) to a steady state (also called quasi-stationary state) for which the grain size distribution function (scaled or normalized with respect to a mean size) is independent of the initial size distribution and time-invariant, i.e., does not change with time any more.

This picture agrees with the idea of statistical self-similarity, which has arisen in early sintering theory (Burke 1949, Mullins 1986).

In two dimensions it is known that in the steady state for a polygon with s sides the mean number of sides of the neighboring polygons $\bar{s}_{neighbor}$ is not arbitrary (in contrast to the case of Voronoi polygons), but a function of s, given by Aboav's law (Aboav 1970, Rivier 1985),

$$\bar{s}_{neighbor}(s) = 5 + \frac{8}{s} . \tag{9}$$

However, a three-dimensional analogue of Aboav's law is not available. Nevertheless, newest research in this field (Zöllner & Rios 2017) has shown that the self-similar coarsening regime is a steady state in which topological transitions (neighbor switching, edge disappearance, grain disappearance) occur as often as required by a law of the von-Neumann-Mullins type, but without changing the spatial distribution of the grains (spatially correlated grain arrangement). This condition is not fulfilled by the Hillert model, which is based on the LSW model.

An enormous influence on the further development of materials science and engineering as a whole, but in particular of ceramic science and engineering, had the "Third International Materials Symposium Ceramic Microstructures – Their Analysis, Significance, and Production," held at the University of California, Berkeley, 13–16 June 1966, the proceedings of which have been published as a book (Fulrath & Pask 1968). This book was the first serious attempt to apply the modern developments of quantitative stereology-based image analysis, which had until then been used mainly in metallography, to ceramic microstructures. This was highly required, because much of the Russian literature on that topic was not available in English at that time. In particular, Saltykov's remarkable book on stereometric metallography, which was at that time available already in a second Russian edition (Saltykov 1958) had to wait another six years for the first and only translation into a Western language (Saltykov 1974). The relative inaccessibility of Saltykov's work is also the reason why some of the stereological relations have been discovered (or rediscovered) several times by different authors. Although stereology has been used increasingly since the 1970s (Exner 2005, Kingery et al. 1976, Russ & DeHoff 2000), its application in materials science and engineering is still not as widespread as would be desirable, and many prejudices and misconceptions still prevail in the literature, see our recent papers on this topic (Pabst & Uhlířová 2017, Pabst et al. 2015).

While traditional work on solid state sintering mechanisms and the associated kinetics was continuing (Exner & Bross 1979, Uskoković & Exner 1977), extended to include the influence of external pressure (Wellner & Exner 1973) and several times summarized (Exner 1980, Exner & Arzt 1983), a new approach that has become popular from the 1970s onwards, following two early key papers by Johnson (Johnson 1969a, 1969b), was the computer simulation of sintering processes (Bross & Exner 1979, Johnson 1990) and grain growth (Srolovitz et al. 1983, 1984, Suh et al. 1996, Fan & Chen 1996). Today, of course, the computer simulation of sintering processes (mainly grain growth, but also densification) has become an enormous field of research, the treatment of which would by far exceed the scope of the present chapter. The interested reader may refer to (Niu & Pan 2010), as well as the papers of Streitenberger and Zöllner (Streitenberger & Zöllner 2006, Zöllner & Streitenberger 2006, Zöllner & Rios 2017) and the references cited therein, for more information.

From the more recent work, probably the most influential single paper in sintering theory in the last 25 years was the 1996 paper by Su and Johnson (Su & Johnson 1996), in which the concept of master sintering curves has been introduced. This concept can be considered as a result of earlier attempts to develop a combined-stage sintering model (Hansen et al. 1992) and turned out to be rather useful for practical purposes, although its theoretical foundation is disputable. This concept is treated in some detail below.

3. Dihedral Angle, Driving Force and Mechanisms of Solid State Sintering

Solid state sintering (SSS) is a process that transforms a powder consisting of small particles at high temperature into a more or less dense solid compact without the help of a liquid (glass) melt. While the primary particles are crystalline, i.e., small single crystals (crystallites), the compact after sintering is a polycrystalline material, in which the crystallites are usually called grains. SSS is always accompanied by strengthening (due to a reduction of free surface area via neck formation, neck growth and particle shape changes) and usually also by densification (due to a reduction of porosity, which is accompanied by shrinkage) and coarsening (due to a reduction of grain boundary area via grain growth).

Generally speaking, the driving force of sintering is the tendency of the system to attain lowest Gibbs free energy (free enthalpy, chemical potential). In SSS this is primarily realized via replacing the free surfaces (i.e., solid-gas or solid-vapor interfaces), which are the interfaces with the highest excess energy (compared to the bulk of the crystallite), by grain boundaries, which have lower energy compared to free surfaces (but still higher energy compared to the bulk). Further reduction of Gibbs free energy is then achieved by reducing the grain boundary density (i.e., the grain boundary area per unit volume of the polycrystalline material) via grain growth.[5]

Strictly speaking, it is the interface excess energy that is reduced by this process, because any type of interface, whether a free surface or a grain boundary, has a higher Gibbs free energy than the bulk of the crystallite. A convenient measure for the ratio between surface free energy per unit area (surface tension) γ_{SV} and grain boundary free energy per unit area (interfacial tension) γ_{SS} is the so-called dihedral angle φ, which is defined via the relation (German 1996, Rahaman 2003)

---

[5] Perfect single crystals have the lowest Gibbs free energy. Compared to single crystals, polycrystalline materials have a higher energy content, which can principally be reduced by infinite grain growth, unless kinetic constraints preclude this.

$$\frac{\gamma_{SS}}{\gamma_{SV}} = 2\cos\frac{\varphi}{2}. \tag{10}$$

Typical surface tensions γ_{SV} for solids below the melting temperature are 1–3 N/m (Rahaman 2003, Stokes & Evans 1997), whereas the interfacial tensions of solid-solid grain boundaries γ_{SS} are typically smaller < 1 N/m (Stokes & Evans 1997). By definition the dihedral angle can principally adopt values ranging from 0 to 180°. A value of $\varphi = 0°$ would correspond to the case $\gamma_{SS} = 2\gamma_{SV}$, which is the maximum value thinkable for γ_{SS}. In this case the driving force for sintering is reduced to zero, i.e., sintering does not occur. Non-zero dihedral angles are required for sintering, and the reason why thermal etching can be used to visualize grain boundaries on polished sections (Munoz et al. 2004, Salmang & Scholze 1982). Indeed, for real materials the dihedral angles are nonzero, and sometimes very high values (φ > 150°) are assumed to occur (Rahaman 2003). However, values close to 180° can occur only when the orientation of grains (crystallites) coincides, in which case a grain boundary is *de facto* absent.[6] Nevertheless, most sintering models are idealized in the sense that they assume φ = 180°, i.e., they assume the grain boundary energy to be negligible compared to the surface energy (i.e., $\gamma_{SS} << \gamma_{SV}$ or $\gamma_{SS} \approx 0$). Only in this idealized case, which is implicitly assumed in most classical sintering models (for all sintering stages), the sintering necks between two particles can have a simple semicircular curvature, and only in this case intergranular pores can be imagined to be cylindrical or spherical. In all other cases the sintering necks have grooves, and the pores can be at best cylinders with non-circular cross sections or polyhedra with curved faces instead of spheres. The situation in real materials is even more complicated due to the fact that the amount of excess grain boundary energy depends on the details of the

[6] This may account for the occasional invisibility of individual grain boundaries in thermally etched polished sections.

interface, for example on the mutual orientation of the two crystallites (Stokes & Evans 1997, Sutton & Balluffi 1995). Therefore, instead of a single dihedral angle, which is an idealization, the dihedral angle in real materials always exhibits a distribution of values. In particular, for magnesia, undoped alumina and MgO-doped alumina the dihedral angles were found to vary in the ranges 89–118°, 75–165° and 90–140°, respectively (Handwerker et al. 1990b), and more recent work (Dillon & Rohrer 2009, Saylor & Rohrer 1999) reported dihedral angles ranges of approximately 70–140°, 60–135°, 70–160° and 135–170° for magnesia, undoped alumina, MgO-doped alumina and yttrium aluminum garnet (YAG), respectively. There are several techniques for measuring dihedral angles, such as, e.g., optical interferometry, calculations from thermal groove width and depth, the so-called MRL (metal reference line) technique (Handwerker et al. 1990a), and AFM-based techniques (Dillon & Rohrer 2009, Saylor & Rohrer 1999). An average value of the dihedral angle can also be obtained by stereology-based image analysis (Russ & DeHoff 2000).

The driving force of sintering can also be considered from another point of view, based on the Young-Laplace equation (Stokes & Evans 1997), which determines the stress associated with curved surfaces (in the context of sintering called “sintering stress”),

$$\sigma = \gamma_{SV} \kappa = \gamma_{SV} \left(\frac{1}{r_1} + \frac{1}{r_2} \right), \tag{11}$$

where κ is the mean curvature and r_1, r_2 are the principal radii of curvature (German 1996, Rahaman 2003, Russ & DeHoff 2000). The sign of this stress value is of course coupled with the sign convention (definition) used for the curvature, resulting in negative stress, i.e., positive (over-) pressure, in convex bumps and solid particles (similar to liquid drops) and positive stress, i.e., negative (under-) pressure, in concave cavities and pores (similar to bubbles in liquids). Note also that the absolute value of the mean curvature is

- $\kappa = \frac{2}{r}$ for spherical objects (with radius $r_1 = r_2 = r$) and
- $\kappa = \frac{1}{r}$ for cylindrical objects (with $r_1 = r$ and $r_2 = \infty$).

Using the Young-Laplace equation, the equilibrium vapor pressure over a curved surface can be calculated via the Kelvin equation,

$$p = p_0 \cdot \exp\left[\frac{\gamma_{SV}\Omega}{kT}\left(\frac{1}{r_1}+\frac{1}{r_2}\right)\right] \approx p_0\left[1+\frac{\gamma_{SV}\Omega}{kT}\left(\frac{1}{r_1}+\frac{1}{r_2}\right)\right], \qquad (12)$$

where p_0 is the equilibrium vapor pressure over a planar surface, Ω the atomic volume, k the Boltzmann constant ($1.381 \cdot 10^{-23}$ J/K), T the absolute temperature and the "$\approx$" sign involves the assumption that the energy term in the nominator is much smaller than that in the denominator (note that stress or pressure corresponds to an energy density, as can be seen from the units [Pa] = [N/m^2] = [J/m^3]), i.e.,

$$\gamma_{SV}\Omega\left(\frac{1}{r_1}+\frac{1}{r_2}\right) << kT. \qquad (13)$$

A Kelvin equation of this type can be (and has been) directly used for the mechanism of evaporation and condensation, which is operative in the sintering of alkali halides, for example (Kingery & Berg 1955). It follows from the Kelvin equation that convex solid particles have a higher vapor pressure than planar solid surfaces of the same material, and on the other hand concave cavities in solids have lower vapor pressure. That means, when convex particles are in contact with a flat surface (Kuczynski 1949) or with each other (Burke 1958, Kingery & Berg 1955), atoms will evaporate from the convex (parts of the) particles and condense in the concave contact regions, forming sintering necks.[7] In other words, concomitantly with the evaporation of atoms from the convex surfaces, which act as atom sources, the concave surfaces, which act as atom sinks, try to reduce their curvature. While the sintering neck grows and the contact area increases in this way, the center-to-center distance of mutually supported particles in powder compacts remains essentially unchanged, because matter from the convex surface, but not from the bulk, is deposited at the concave surface. That means, the evaporation-condensation mechanism leads to particle shape changes and an increase in the contact area that results in a strengthening of the powder compact (sintered body) without densification and thus without shrinkage. In other words, if evaporation-condensation was the only sintering mechanism operating, the sintered body would have the same porosity as the green body before sintering. However, imagining a one-dimensional chain of particles, the final result of sintering would not necessarily be a cylindrical wire (with spherical end caps and a diameter smaller than the initial particles), but usually a "chain of beads," because although the driving force drives the system toward zero mean curvature, this does not necessarily mean planar surfaces. Experimental evidence of this effect has been called "de-sintering" and is found when heating polycrystalline fibers (Lange 1996, Miller & Lange 1989). Saddle surfaces, albeit curved, can have zero mean curvature, when the convex and concave surface patches compensate each other. Moreover, as mentioned above, cylindrical surfaces, even if straight in the axial direction and thus exhibiting lower mean curvature than spheres, are still convex in the transverse direction. That means, the final microstructure of a sintered powder compact in which only the non-densifying evaporation-condensation mechanism was operative may be quite porous, consisting of an appropriate combination of saddle surfaces, instead of a dense polycrystalline material with closely polyhedral fitting grains. Nevertheless, in a similar manner as the driving force of sintering aims at reducing the surface curvature of particles by neck growth, it also

[7] More precisely, a sinter neck is a saddle surface, convex in the contact plane and concave in the planes perpendicular to the contact plane. However, the absolute value of concave curvature is much higher than the magnitude of convex curvature, so that typical sinter necks can be considered as being dominated by concave surfaces.

aims at grain growth, because smaller grains have a higher curvature and thus higher vapor pressure than large grains.

Now it can be argued that sintering via the evaporation-condensation mechanism is not solid state sintering (SSS) *sensu stricto*, since a vapor phase is involved that is responsible for mass transport. In fact, the evaporation-condensation mechanism has been confirmed mainly for halides and might be expected to be significant or dominant only for a few oxides with high vapor pressure such as ZnO or PbO. While the viscous flow mechanism is known to be operative in amorphous powders (glasses) and plastic deformation mainly in metal powders,[8] it was one of the key findings of early sintering theory (Burke 1958, Kingery & Berg 1955, Kuczynski 1949) that the diffusion of vacancies is the basic SSS mechanism in most oxide ceramics. In this case the driving force is not the difference (or gradient) in vapor pressure but the difference (or gradient) of the vacancy concentration. An equation completely analogous to the Kelvin equation for the equilibrium vapor pressure can be used to describe the equilibrium vacancy concentration under a curved surface, viz.

$$C = C_0 \cdot \exp\left[-\frac{\gamma_{SV}\Omega}{kT}\left(\frac{1}{r_1}+\frac{1}{r_2}\right)\right] \approx C_0\left[1-\frac{\gamma_{SV}\Omega}{kT}\left(\frac{1}{r_1}+\frac{1}{r_2}\right)\right], \quad (14)$$

where C_0 is the equilibrium vacancy concentration under a planar surface and the approximation on the right hand side of this equation holds under the same conditions as before, i.e., under the assumption that the energy term in the nominator is much smaller than that in the denominator. That means, in the vicinity of a concavely curved surface (surface cavity), for example in the solid shell surrounding a cylindrical pore channel (with mean curvature $\kappa = 1/r$) or spherical pore (with mean curvature $\kappa = 2/r$), the vacancy concentration is higher by

[8] However, high-temperature plastic deformation or creep in pressure-assisted sintering cannot be excluded even in ceramics, as long as slip planes are available and dislocation movement is possible.

$$\frac{\Delta C}{C_0} = \frac{\kappa \, \gamma_{SV} \Omega}{kT}, \tag{15}$$

compared to that under a planar surface. Principally it is this difference (or gradient) in the concentration of vacancies that effectively leads to the counterflow of matter (atoms) in order to counterbalance this difference (gradient).[9] Early sintering theory (Kuczynski 1949) has clearly shown, however, that the mechanism is usually realized by the movement of vacancies, not atoms (e.g., the interstitial or exchange mechanism), see (Rahaman 2003). From a phenomenological point of view this movement of vacancies, like that of any other atomistic-scale species, can be described by Fick's law of diffusion (Crank 1975),

$$\mathbf{j} = -D \, \mathrm{grad}\, C, \tag{16}$$

where $\mathbf{j}$ is the mass flux vector, $\mathrm{grad}\, C$ the concentration gradient (here of vacancies) and D the diffusion coefficient (of vacancies). It is intuitively clear that diffusion is easier when the diffusion path is more accessible, i.e., less dense and less rigid. Since the vacancy concentration is generally higher at the surface than in the bulk of particles or grains (crystallites), it can be expected that diffusion at and along the surface is easier than through the bulk. This was well known already in early sintering theory (Burke 1958, Kingery & Berg 1955, Kuczynski 1949). Both Burke (Burke 1958) and Coble (Coble 1961a) emphasized the significance of grain boundaries and pointed to the fact that grain boundaries are usually less dense than the bulk of the grains (crystallites), see also (Pabst and Gregorová 2010, Stokes & Evans 1997, Sutton & Balluffi 1995). Coble identified pores as vacancy sources (formation sites) and grain boundaries as vacancy sinks (annihilation sites) (Coble 1961a).[10]

---

[9] At the same time the concentration of vacancies is higher at the surface and in the grain boundary than in the bulk.

[10] Since tilt boundaries can be considered as being comprised of edge dislocations (Coble 1961a, Stokes & Evans 1997, Sutton & Balluffi 1995), one may also say that dislocations act as

Therefore it makes sense to distinguish surface diffusion, grain boundary diffusion and volume diffusion. While the first of these mechanisms is non-densifying, the second and usually also the third[11] lead to densification (and thus also to shrinkage). Diffusion coefficients are strongly temperature-dependent and usually exhibit an exponential increase with increasing temperature. Since diffusion can be considered as a thermally activated process, the temperature dependence of the diffusion coefficients can be described by an Arrhenius equation of the type

$$D = D_0 \cdot \exp\left(-\frac{Q}{RT} \right), \tag{17}$$

where the nominator Q in the exponent can be interpreted as an activation energy and R is the universal gas constant (8.314 J/molK). Now the activation energy for surface diffusion is lowest, that of grain boundary diffusion intermediate and that of volume diffusion highest. Therefore surface diffusion will be more prominent at low temperatures, grain boundary diffusion at intermediate temperatures and volume diffusion at high temperatures. A thinkable mechanism that is commonly ignored so far is triple junction diffusion.

Although the initial stage neck growth kinetics (see below) can in principle be used to identify the sintering mechanism (Exner & Arzt 1983), this task is rather difficult, especially when the grain boundary energy is high (and thus the dihedral angle small), see (German 1996, Rahaman 2003). Moreover, in practice several sintering mechanisms can operate simultaneously. The so-called sintering maps proposed by Ashby (Ashby 1974, Swinkels & Ashby 1981) can be useful tools for a qualitative

---

vacancy sinks. On the other hand, grain boundaries can also be sources of vacancies (German 1996, Rahaman 2003).

[11] Volume diffusion, also called bulk or lattice diffusion, usually leads to densification (and shrinkage), but in principle also non-densifying volume diffusion (diffusion through the lattice) is thinkable, when the free surface acts as a vacancy source instead of the grain boundary (German 1996, Rahaman 2003). It may be expected that the latter effect will be more significant in nanocrystalline materials, because there the surface density is higher than in coarse-grained materials.

assessment of their relative contributions and the expected trends in dependence of the sintering temperature and heating rate. However, in quantitative terms, their predictive power of sintering maps is rather limited, because they are based on the classical (two-sphere) models, and small changes in the input parameters can significantly change the sintering map, see (Uskoković & Exner 1977, Rahaman 2003).

Since stresses can be considered as negative pressures, the sintering stress (Rahaman 2003, 2010) calculated from the Kelvin equation (for sintering at atmospheric pressure) can be interpreted as if an external pressure was acting on the powder compact. Therefore the driving force of sintering can be increased by additional external pressure, which can be uniaxial (hot pressing/HP, spark plasma sintering/SPS) or isostatic (hot isostatic pressing/HIP). In this case high-temperature plastic deformation (creep) comes into play, and hot pressing models are available that combine the classical SSS models with creep models for volume diffusion, so-called Nabarro-Herring creep (Herring 1950b, Nabarro 1948), or grain boundary diffusion, so-called Coble creep (Coble 1963). This topic, however, is beyond the scope of the present chapter. For further information the reader may refer to (German 1996, Rahaman 2003, 2010) and the references cited therein. For the special topic of electric-current assisted pressure sintering, also called spark plasma sintering (SPS), electric current activated/assisted sintering (ECAS) or field activated sintering technique (FAST), see also (Munir et al. 2006, Orru et al. 2009, Quach et al. 2010). Excellent summaries are also available for other sintering techniques, such as controlled atmosphere sintering (Blais 2010), vacuum sintering (Heaney 2010), microwave sintering (Agrawal 2010), and photonic sintering (West et al. 2010). It should be mentioned that most of these sintering techniques can be applied both to ceramics and to metals, for example to aluminum and its alloys (Qian & Schaffer 2010), titanium and its alloys (Qian et al. 2010), and refractory metals (Johnson 2010). Very special sintering procedures have been applied to ultrahard materials, such as diamond and boron nitride (Belnap 2010).

4. STAGES AND KINETICS OF SOLID STATE SINTERING

The solid state sintering (SSS) process can be divided into three stages, *viz.* initial, intermediate and final. While early sintering theory was very much focused on neck growth in the initial stage (Kingery & Berg 1955, Kuczynski 1949) or grain growth in the final stage of sintering (Burke 1958, Burke & Turnbull 1952), starting with Coble's pioneering work (Coble 1961a) the intermediate stage of SSS was clearly defined and recognized as the most important stage for densification.[12] A crucial step in Coble's work was his choice of the tetrakaidecahedron (plane-faced Kelvin cell) as a model shape for grains in polycrystalline materials, see Figure 1. Indeed this choice is a very useful one, because the Kelvin cell is a space-filling shape that accounts for the fact that polyhedral grains in dense polycrystalline materials compacted from randomly arranged spherical particles have 14 faces on average, not 12 (Exner 1979, German 1996).[13] These realistic features are exhibited neither by pentagon dodecahedra nor by rhombic dodecahedra (Kepler cells), see (Aste & Weaire 2000). On the other hand, Scherer's choice of a cubic cell for modeling intermediate-state sintering of amorphous materials (Scherer 1977a, 1977b, 1984) has the advantage of being computationally simpler.

Based on the tetrakaidecahedral grain shape model for intermediate and final stages, together with the two-sphere model for the initial stage, Coble was able not only to explain qualitatively the processes and topological changes occurring during the individual sintering stages, but also to perform quantitative calculations of the sintering kinetics, see also (Rahaman 2003). This has been possible because the Kelvin cell is a well defined model shape, for which the number of corners (vertices), edges and faces is known and the edge length, surface area and volume can be easily calculated. However, the most important feature of this – admittedly highly

---

[12] Very roughly it could be said that the initial stage of SSS is responsible mainly for strengthening, the intermediate stage mainly for densification and the final stage mainly for grain growth.

[13] This fact has long before been recognized in the context of biological tissues (Lewis 1950) and foams, e.g., soap froth (Matzke 1946), see (Aste & Weaire 2000). Rhombic dodecahedra, so-called Kepler cells, are observed only when the initial packing is ordered (Marvin 1939).

idealized – model is that the major topological changes during sintering are correctly displayed. From the viewpoint of current knowledge the stages of SSS can be described as follows.

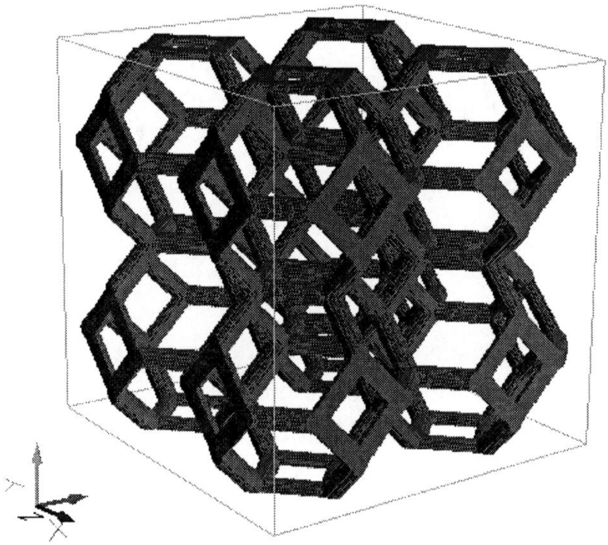

Figure 1. Schematic view of a space-filling assemblage of tetrakaidecahedral cells (plane-faced Kelvin cells), one of the simplest, but quite realistic, model geometries for dense polycrystalline materials; the struts in this figure may be imagined as representing an interconnected network of cylindrical pore channels, corresponding to Coble's model for the intermediate stage of sintering (in the final stage this cylindrical pore channel network would break up into an array of spherical pores situated at the nodes of this network, i.e., at the cell corners or vertices).

Initial Stage

The initial stage of SSS usually starts with powder green compacts in which the particles are in dense random packing (maximally random jammed packing), corresponding to a relative density (solid volume fraction) of approximately 64% in the case of monosized rigid spheres (German 1996, 2010, Torquato 2002, Torquato et al. 2000). In real systems, of course, the particles are usually neither spherical nor monosized, so that considerably smaller initial relative densities are usually assumed or reported for real materials, e.g., 0.5–0.6 (Rahaman 2003). During the initial phase of SSS the point contacts between the convex particles are transformed into areal contacts (necks), thus strengthening the powder compact (network of particles) and forming a solid skeleton connected via sinter necks. In this initial stage the particle shape starts to change, before finally (in the later stages of sintering) particles of any initial shape are transformed into polyhedral grains. Depending on the mechanism of neck growth, this strengthening may or may not be accompanied by densification and shrinkage. In particular, if isothermal SSS is performed at low temperature, surface diffusion may be the most important mechanism. In this case, given sufficiently long time, strengthening may principally occur without significant densification and shrinkage and thus result in an ideally bicontinuous microstructure (Pabst et al. 2016).[14] Another non-densifying SSS mechanism that can play a role in the initial stage of SSS for materials with high vapor pressure is evaporation-condensation, whereas volume (lattice) diffusion with vacancy sources at the surface, albeit also non-densifying (German 1996, Kingery et al. 1976, Rahaman 2003, 2010), is a mechanism that requires high activation energy and will compete with surface diffusion only at higher temperatures. At high temperatures, however, the densifying mechanisms, i.e., grain boundary diffusion and volume (lattice) diffusion with vacancy sources at the grain boundary (German 1996, Kingery et al. 1976, Rahaman 2003, 2010), are operating in addition to the non-densifying mechanisms, so that shrinkage will occur anyway. Initial-stage SSS is usually defined for neck-to-grain size ratios X/G smaller than 0.3 (Frenkel 1945, Kuczynski 1949, Kingery & Berg 1955, German 1996), but Rahaman assumes a considerably higher limiting value of 0.4–0.5

---

[14] In practice this is encountered only occasionally. It does, however, occur in tin oxide ceramics (Gouvea et al. 1998, Fayette et al. 2000, Poulier et al. 2007) and could occur also in other systems with high vapor pressure, such as zinc oxide. Moreover, it seems to occur quite commonly also in nanocrystalline systems, where surface diffusion is the dominant sintering mechanism (Fang & Wang 2010). Already in 1970 Procházka and Coble (Procházka and Coble 1970) have reported, that a 0.2 μm alumina powder sintered at 750 °C for times ranging from a few minutes to several days, exhibited surface reduction kinetics typical for surface diffusion, see (German 1996).

(Rahaman 2003).[15] There is similar disagreement about the maximum shrinkage during the initial stage. While German admits a shrinkage of maximally 3% and an increase in relative density from 64% to 70% during the initial stage of SSS (German 1996, 2010), Rahaman admits shrinkage values up to 5% and a relative density increase from 40–50% to 65% (Rahaman 2003). The most popular and (because of its simplicity) probably most useful model for analytical calculations of neck growth and its kinetics is the two-sphere model (Kingery & Berg 1955), see also (German 1996, 2010, Kingery et al. 1976, Rahaman 2003, 2010). It has to be emphasized, however, that this model is idealized not only because spherical particles are considered or because only two (and not more) such particles are considered[16] or because both particles are usually assumed to have the same size, but also because the concave neck radius is considered to be smooth (semicircular), which is possible only for a dihedral angle of 180°. In reality the necks exhibit grooves, a fact that is expected to change the sintering kinetics (German 1996). Therefore the model equations for initial-stage neck growth have to be considered as generally oversimplified.

Intermediate Stage

As soon as the free surfaces have been largely replaced by grain boundaries, SSS enters the intermediate stage. In this stage the microstructure is bicontinuous (Pabst et al. 2016) and, while the grains are connected via growing contact areas, the pore space still remains fully interconnected. In this stage the pore space can be considered as a network of pore channels along the developing grain edges. Densifying mechanisms usually prevail in this stage (Rahaman 2010), so that the porosity is significantly reduced, concomitantly with the pore size (i.e., the pore

---

[15] It should be noted, that Rahaman's definition is in serious conflict with German's, because the latter considers the neck-to-grain size range between 1/3 and 1/2 as typical for intermediate stage sintering (German 2010).

[16] Actually, multisphere models have also been proposed and used (Pask 1987).

channel cross section area). Thus most of the total shrinkage occurs in this stage, and the relative density increases from 65–70% to 90–92% (German 2010, Rahaman 2003). It should be noted that according to German (German 2010), but in contrast to Rahaman's aforementioned definition of initial state sintering (Rahaman 2003), the neck-to-grain size ratio during intermediate-stage SSS is between 1/3 and 1/2. According to Coble's idealized microstructural model for the intermediate stage of SSS (Coble 1961a) the microstructure in this stage consists of tetrakaidecahedral grains with cylindrical pore channels along the edges, see Figure 1, forming an interconnected network. The realistic feature of this model is that the interconnected pore channel network reflects very well the completely open porosity that prevails during the intermediate stage of SSS, as well as the bicontinuous character of the microstructure, which is typical for partially sintered materials (Pabst et al. 2016). Less realistic is the assumption of cylindrical pores with circular cross sections, because this assumption implies that the grain boundary energy is negligible compared to the surface energy, which is usually not the case (see above).[17] As a consequence of the non-negligibility of the grain boundary energy and the associated fact that the dihedral angles in real polycrystalline materials is considerably smaller than 180°, the cross section of the pore channels in real materials will have the shape of a circularly inflated triangle.[18] It is the competition between this curvature (and the associated tendency to eliminate it by grain boundary movement toward the center of curvature) and the equilibrium dihedral angle that governs the thinning kinetics of the pore channels and thus the densification rate.

---

[17] This assumption is strictly justified only if the grain boundary is *de facto* absent, i.e., when two crystallites have the same orientation or if the grains are amorphous. Actually, for glasses this condition is fulfilled. Thus, paradoxically, Coble's model (Coble 1961a), which is intended for (and has always been applied to) polycrystalline materials (Coble 1961b, German 1996, Rahaman 2003), contains an assumption that holds strictly only for glasses. In contrast to Coble's model for intermediate-stage sintering of polycrystalline materials, Scherer's model for intermediate-stage sintering of amorphous materials (Scherer 1977a, 1977b, 1984) is based on a cubic grain shape model.

[18] Note that the stereological evaluation of planar sections through these pore channels can be used to measure an average dihedral angle (Russ & DeHoff 2000).

Final Stage

The intermediate stage ends when the porosity is around 8–10% (German 2010, Rahaman 2003), which is the value at which the pore channels become too thin to be stable. Actually it has been stated that, independent of the initial density, the pore structure exhibits a transition from open to closed porosities at total porosities between 12% and 6%, but the dihedral angle determines the position and the initial particle and pore size distribution the sharpness of this pore closure transition (German 1996). The situation can be (and has been) compared with the Rayleigh instability in a liquid jet (Rahaman 2003), which disintegrates into individual droplets when the jet diameter becomes too small, as a consequence of surface tension (because a sphere has a lower surface energy than a cylinder of the same volume). As a result the pores become isolated and the microstructure type changes from bicontinuous to matrix-inclusion (Pabst et al. 2016). While the open porosity exhibits a steep decrease, the percentage of closed porosity exhibits a steep increase and goes through a maximum in the vicinity of the point of pore closure, which is usually somewhere in the range 6–12% porosity (German 1996, Thummler & Thomma 1967), see Figure 2. After passing this maximum at around 6.5% porosity (German 1996) also the closed porosity decreases, albeit with a slower rate than the open porosity (Coleman & Beere 1975, Rahaman 2003). According to Coble's idealized Kelvin cell model (Coble 1961a) the closed pores in the final stage of SSS are spherical and situated at the corners (vertices, quadruple nodes) of the tetrakaidecahedral grains. In reality, however, these closed pores are usually neither spherical nor necessarily situated at the grain corners. In cases where they happen to be situated at the grain corners, however, the closed pores tend to approach the shape of a spherically inflated tetrahedron instead of a sphere, because the dihedral angle is smaller than 180°. On the other hand, when situated along the grain edges they tend to adopt an elongated (prolate sharp-tipped) and on the grain boundaries a flattened (oblate lenticular) shape, see (German 1996). Since in the final stage of SSS grain growth (see below) competes with pore elimination, it can happen that the grain boundary mobility is higher than the pore mobility, so that the grain boundary "overrides" the pore. Once stranded inside a grain (crystallite), the pore shape will become isometric again, adopting a shape that is either close to spherical or dictated by the crystallographic structure of the bulk crystallite. At this point the pores cannot be eliminated any more within practical sintering times, because volume (lattice) diffusion with its very high activation energy is the only mechanism operating.

During initial-stage SSS the isothermal neck growth kinetics for two-sphere models can be described by the general relation

$$\left(\frac{X}{G}\right)^m = B \cdot \frac{t}{G^n}, \tag{18}$$

where X is the neck diameter, G the particle or grain diameter, X/G the neck size ratio (neck-to-grain size ratio), t the (isothermal) sintering time, and constant B is exponentially dependent on the (absolute) temperature T, according to the Arrhenius-type equation

$$B = B_0 \cdot \exp\left(-\frac{Q}{kT}\right). \tag{19}$$

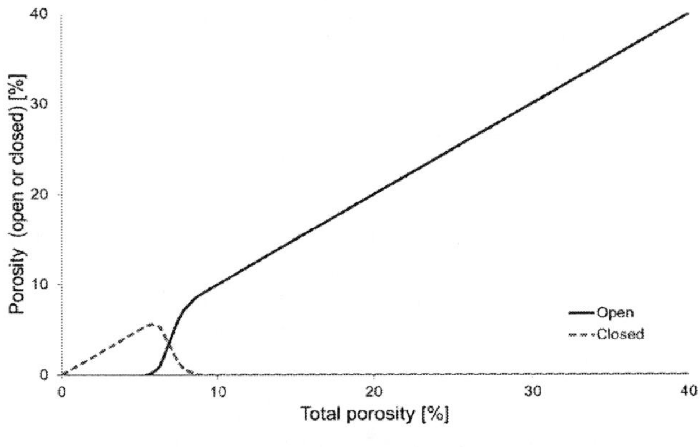

Figure 2. Schematic view of the transition from open to closed porosity at the end of the intermediate and the beginning of the final stage of solid state sintering.

The pre-exponential factor B_0 is a function of geometrical and material parameters, like surface energy γ_{SV}, vacancy volume Ω, surface diffusion layer thickness δ_S, grain boundary diffusion layer thickness δ_{GB}, which are both of order 1 nm (Rahaman 2003), while the exponential temperature dependence is typical for the diffusion coefficients of surface (D_S), grain boundary (D_{GB}) and volume diffusion (D_V), with $Q_S < Q_{GB} < Q_V$, the solid vapor pressure p_0 that determines the evaporation-condensation mechanism, and (with opposite sign in the exponent) the viscosity η that determines the viscous flow mechanism. All these mechanisms are listed in Table 2 for the sake of completeness and reference. The neck size ratio exponent m and grain size exponent n exhibit characteristic values depending on the prevailing sintering mechanism, see Table 2.

Table 2. Theoretical values of initial-stage neck size exponent m and grain size exponent n for different sintering mechanisms (Exner 1979, Exner & Arzt 1990, German 1996, Rahaman 2003, 2010)

Sintering mechanism	m	n	Factor $B \propto$	Densification
Surface diffusion	7	4	$\frac{D_S \delta_S \gamma_{SV} \Omega}{kT}$	No
Grain boundary diffusion	6	4	$\frac{D_{GB} \delta_{GB} \gamma_{SV} \Omega}{kT}$	Yes
Volume diffusion from the grain boundary	5	3	$\frac{D_V \gamma_{SV} \Omega}{kT}$	Yes
Volume diffusion from the surface	4	3	$\frac{D_V \gamma_{SV} \Omega}{kT}$	No
Evaporation-condensation (vapor transport)	3	2	$\frac{p_0 \gamma_{SV} \Omega}{(kT)^{3/2}}$	No
Viscous flow	2	1	$\frac{\gamma_{SV}}{\eta}$	Yes

These values, which have been calculated from idealized models, have been called "plausible" (Exner & Arzt 1990, Rahaman 2010). Actually, in practice it is not easy to unambiguously determine the sintering mechanism from kinetic measurements, because several mechanisms can act simultaneously (Ashby 1974, Rahaman 2003).[19] In particular, according to Exner and Arzt (Exner & Arzt 1990) the experimentally measured values of the exponents for surface diffusion can be as low as those for evaporation-condensation, while those for evaporation-condensation can be as high as those for surface diffusion. With regard to this finding one may assume that the two mechanisms often act simultaneously and cannot be easily distinguished. Also it is interesting that the grain size exponent n occurring in the initial-stage neck growth kinetics, Equation 18, has the same values as the exponent M in Herring's scaling law, see Equation 4.

Important qualitative conclusions can be drawn from these initial-stage relations. A high sensitivity to the inverse particle or grain size means that smaller particle size (grain size) results in faster sintering (i.e., neck growth = strengthening, and possibly densification). Since surface and grain boundary diffusion have the highest grain size sensitivity, these mechanisms are especially favored (compared to the other mechanisms) by a smaller particle size. Note that in all cases temperature enters these relations via an exponential expression, meaning that small temperature changes have a large effect on the neck growth kinetics. The sintering time has a relatively weak effect in comparison to both temperature and particle size. Non-isothermal sintering conditions due to a high heating rate, for example, can have a more significant effect on initial-stage sintering than intermediate-stage and final-stage sintering.

In the intermediate and final stages of SSS sintering kinetics are usually described via empirical equations, which serve as fit models. For polycrystalline materials the modeling of both stages can be based on the Coble model (Coble 1961a), i.e., a microstructure consisting of tetrakaidecahedral grains with open cylindrical pores along the edges in the

---

[19] Therefore it makes no sense to cite the numerical coefficients occurring in these relations, which follow exclusively from the special geometry assumed, usually the two-sphere model.

intermediate stage and closed spherical pores at the corners in the final stage. Since this model assumes uniform and unchanged pore shape, non-densifying mechanisms cannot operate and therefore only grain boundary and lattice diffusion are considered in the intermediate- and final-stage sintering of polycrystalline materials.[20] Based on the Coble model (Coble 1961a), Rahaman (Rahaman 2003, 2010) calculated the densification rates for volume (lattice) diffusion as

$$\frac{d\rho_r}{dt} \propto \frac{D_V \gamma_{SV} \Omega}{kT} \cdot \frac{1}{G^3} \tag{20}$$

and for grain boundary diffusion as

$$\frac{d\rho_r}{dt} \propto \frac{D_{GB} \delta_{GB} \gamma_{SV} \Omega}{kT} \cdot \frac{1}{G^4} . \tag{21}$$

In these relations $d\rho_r/dt$ is the time derivative of the relative density or volume fraction of solid phase, which is related to the porosity ϕ via the relation

$$\rho_r = \frac{\rho}{\rho_0} = 1 - \phi , \tag{22}$$

where ρ is the bulk density and ρ_0 the theoretical density. Note that – except for numerical factors (which are not considered here, because they are only valid for the special geometry of Coble's model) and a density-dependent correction term, the stress intensification factor (Rahaman 2003, 2010) – these relations are formally identical for the intermediate and final

[20] In amorphous materials only the viscous flow mechanism has to be taken into account – the intermediate stage is in this case described by the Scherer model, which assumes cubic grains with cylindrical pores along the edges (Scherer 1977a, 1977b, 1984), the final stage by the Mackenzie-Shuttleworth model (Mackenzie & Shuttleworth 1949), see also (Hlaváč 1988, Rahaman 2003). Sintering of amorphous systems is not considered in this chapter.

stages of sintering. Moreover, with respect to the fact that the surface tension multiplied by the curvature can be interpreted as a sintering stress (German 1996, 2010), these equations are similar to the creep equations and the hot pressing equations, which play an important role in pressure-assisted sintering (Rahaman 2003, 2010). Interestingly, the grain size exponents in these relations agree with those for initial-stage SSS, and it is clear that the densification rate decreases with increasing grain size. Contrary to the external pressure, which adds to the curvature-induced sintering stress and thus helps densification by increasing the densification rate during hot pressing/HP and hot isostatic pressing/HIP without affecting the grain growth, in the final stage of SSS the internal gas pressure of closed pores reduces the sintering rate and thus impedes full densification, especially if the gas is insoluble in the solid (Markworth 1972). This gas pressure can even lead to the growth of pores by Ostwald ripening of the pore population, and in extreme cases also to an increase of porosity, in the final stage of SSS (Markworth 1973). Therefore the best method to achieve full density is vacuum sintering or at least sintering in the atmosphere of a gas that is soluble in the solid (in oxides, e.g., oxygen or hydrogen, but not nitrogen or argon), see also (German 1996, Salmang & Scholze 1982).

Unfortunately the kinetics of real SSS processes is usually too complex to be satisfactorily described by this model. Therefore empirical relations have been proposed and are often used to fit observed densification kinetics, for example the semi-logarithmic relation (Coble 1961a)

$$\rho_r = \rho_{r0} + K \ln\left(\frac{t}{t_0}\right), \tag{23}$$

where $t - t_0$ is the sintering time, ρ_{r0} the initial relative density at the initial time t_0 and K an empirical rate constant.[21] Of course,

---

[21] A completely analogous relation can be formulated for the time derivative of the relative density (German 1996).

concomitantly with the density increase, the porosity decreases according to the relation

$$\phi = \phi_0 - K \ln\left(\frac{t}{t_0}\right). \qquad (24)$$

Many other empirical fit relations can be found in the literature (Exner 1980, Rahaman 2003). In our opinion a Weibull-type function would be an adequate fit relation for this purpose.

There are many methods to record SSS kinetics. As long as densifying mechanisms are dominant, it is common practice to record the linear shrinkage, either *in situ* or *ex situ*, as a function of the sintering temperature and time. In the initial stage this linear shrinkage can be related to the neck-to grain size ratio X/G via the aforementioned approximation (Burke 1958, Kingery & Berg 1955)

$$\frac{\Delta L}{L_0} = \frac{h}{G} \approx \left(\frac{X}{2G}\right)^2, \qquad (25)$$

where h is half the interpenetration distance of the spheres, see also (Exner 1979, Exner & Petzow 1975, German 1996, Hlaváč 1988, Rahaman 2003, Uskoković & Exner 1977). This leads to the following relation describing the shrinkage kinetics in the initial stage:

$$\left(\frac{\Delta L}{L_0}\right)^{\frac{n}{2}} = B \cdot \frac{t}{2^n G^m}. \qquad (26)$$

In the case where non-densifying sintering mechanisms (e.g., surface diffusion, evaporation-condensation or lattice diffusion with vacancy source at the surface) prevail or shrinkage is completely absent, changes in the free surface area can be a useful measure of the degree of sintering, so that the relation (German 1996, German & Munir 1976)

$$\left(\frac{\Delta S}{S_0}\right)^{\frac{n}{2}} \propto t \tag{27}$$

is useful for extracting information on the sintering mechanism. Further methods are discussed in (Exner 1980, Exner & Arzt 1983). It has to be emphasized that in principle also changes of physical properties can be used to monitor sintering processes, including initial stage sintering. As an interesting example, Figure 3 shows the evolution of Young's modulus for an alumina ceramic body that has been prefired (bisque-fired) to 1050 °C (heating rate 2 °C/min, dwell time 2 h) and has subsequently (after cooling) been submitted to a temperature-dependent impulse excitation measurement (ASTM E 1876-99). That means the bisque-fired ceramic was heated up to 1500 °C (with a heating rate of 5 °C/min, without dwell at maximum temperature), followed immediately by controlled (5 °C/min) cooling to room temperature. The elastic modulus has been measured *in situ* both during heating and cooling.

During this heating and cooling cycle, sintering proceeds, and the porosity decreases from initially 37.8% to 15.6% after the heating-cooling cycle. Thus the curve describes a sintering cycle from a highly porous bisque-fired body (with 37.8% porosity and a very low Young modulus, which is around 25 GPa at room temperature) to a less porous alumina ceramic (with 15.6% porosity and a much higher Young modulus, which is around 260 GPa at room temperature). It is evident that upon heating the Young modulus decreases slightly, as long as the temperature is below approximately 1200°C. This is in perfect agreement with what is expected for fine-grained alumina ceramics (Pabst et al. 2013). However, above that temperature Young's modulus exhibits a steep increase, indicating the continuation of sintering.[22] After exceeding the maximum temperature Young's modulus exhibits a further increase upon cooling, as expected (Pabst et al. 2013). This example shows that sintering processes can be conveniently monitored *in situ* via the evolution of Young's modulus using the impulse excitation technique. It should be emphasized that in contrast to dilatometric measurements this technique is able to monitor strengthening during sintering with or without densification.[23]

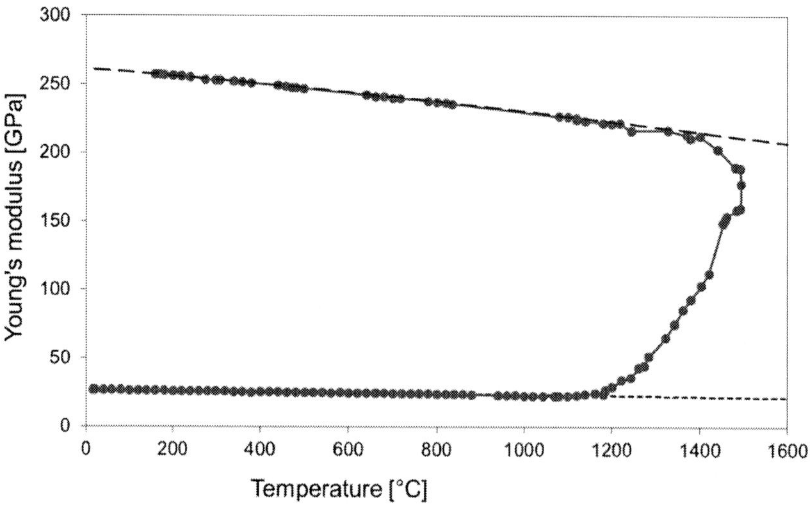

Figure 3. Evolution of Young's modulus (measured via impulse excitation) with temperature for a partially sintered (1050 °C, heating rate 2°C/min, dwell time 2 h) alumina ceramics heated (lower curve with finely dashed fit curve) to 1500°C (heating rate 5°/min, no dwell at maximum temperature) and subsequently cooled (upper curve with coarsely dashed fit curve); from (Gregorová et al. 2012).

---

[22] The reason why the sintering does not continue exactly at the previous bisque-firing temperature is the reduced sintering activity of the grains in the powder compact after the first firing, a phenomenon commonly encountered in practice.

[23] A similar technique has also been used in archeological context to probe the firing temperature of archeological ceramics (potsherds), see (Heimann 1978, Heimann 1982, Heimann & Franklin 1981).

Pore shrinkage generally reduces the free surface area, but increases the surface energy, because of the increased curvature (Rahaman 2003). Moreover, when a grain boundary intersects a pore, a groove will appear according to the equilibrium dihedral angle and affect the sintering rates (German 1996). Therefore pore shrinkage or growth during the intermediate and final stages of SSS is controlled by a complex interplay between the pore size (and thus the number of grains surrounding a pore) and the dihedral angle. For a given dihedral angle there is a critical pore size (or number of grains surrounding the pore), for which the pores are stable, i.e., neither shrink nor grow (Kingery et al. 1976). In general, pores tend to shrink either if they are small or if the grain boundary energy of the surrounding grains is low. In more quantitative terms the conditions for pore shrinkage, stability and growth can be read off from the well-known graph given by Kingery (Kingery et al. 1976), which is reproduced in most textbooks of sintering (German 1996, Rahaman 2003). For example, for a dihedral angle of 120° pores with less than 12 surrounding grains will shrink, while pores with more than 12 surrounding grains will grow. On the other hand a pore surrounded by only 4 grains will shrink even for dihedral angles as low as 60°, while a pore surrounded by 30 grains will not shrink even for dihedral angle as high as 160°. Moreover, the evolution of the porous microstructure is related in a complex way to the evolution of the grain microstructure (Rahaman 2003) and the sintering atmosphere (Blais 2010, German 1996, Salmang & Scholze 1982).

5. Grain Growth Models and Grain Size Distributions

Apart from strengthening, i.e., neck growth (already in the initial stage) and densification, i.e., porosity reduction (mainly in the intermediate stage), SSS is usually accompanied by coarsening, i.e., grain growth (starting in the intermediate stage, but occurring mainly in the final stage). Since the early treatments of grain growth in polycrystalline microstructures (Beck et al. 1948, Burke & Turnbull 1952) it has been assumed that grain growth obeys a relation of the form (Atkinson 1988, Hillert 1965, Louat 1974)

$$G^{\gamma} - G_0^{\gamma} = Ct\,, \tag{28}$$

where G is a mean grain size at time t, G_0 the initial mean grain size and C a rate constant. According to the classical Burke-Turnbull model (Burke & Turnbull 1952) the grain growth exponent γ should be 2, resulting in so-called parabolic or square-root-of-time grain growth kinetics. This theoretical result, based on the assumption that grain growth is a curvature-driven process and is realized via the migration of grain boundaries towards their centers of curvature (realized by atom transport across the boundary), has later been confirmed by other analytical models (Mullins & Vinals 1989), computer simulations (Grest et al. 1988, Weaire & Kermode 1984, Zöllner 2006), and some experiments (Glazier et al. 1987, Weaire & Kermode 1983). However, by far not all experiments support a value of 2. The experiment-based grain growth exponent is usually reported to adopt values in the range 2–4 (Atkinson 1988, Brook 1976, Rahaman 2003), with the case $\gamma = 3$ being most commonly observed in ceramics (so-called cubic or cube-root-of-time grain growth kinetics). It is usually assumed that deviations from the parabolic grain growth kinetics are due to the fact that interactions with dopants, second-phase inclusions and pores have not been taken properly into account in the theoretical model. It has to be noted, however, that also the role of triple junctions and quadruple nodes and topological constraints (von-Neumann-Mullins law, Aboav-Weaire law) are usually not taken into due account, with the exception of a few recent papers (Streitenberger & Zöllner 2015, Zöllner 2013, Zöllner & Rios 2017). Also these could be sources of deviations from the ideal parabolic growth law. The constant C has been shown to vary with grain shape and grain size distribution (Han & Kim 1995, Louis & Gokhale 1995).

The driving force of grain growth is of course the reduction of grain boundary area and the associated interfacial energy. Since both mass and volume conservation must be fulfilled in SSS, large grains grow at the expense of small ones. In a single-phase polycrystalline microstructure this process, which corresponds to Ostwald ripening in two-phase systems (Ratke & Voorhees 2002), is accompanied by the migration of grain boundaries toward their centers of curvature, with a rate proportional to the curvature itself (Burke 1958). Now, apart from the fact that the mean grain size in the microstructure of polycrystalline materials increases during sintering, also the grain size distribution changes. It is common practice to distinguish normal (or continuous) or abnormal (or discontinuous) grain growth.[24] The latter can occur, e.g., when the initial distribution is bimodal, i.e., when a relatively small amount of large particles is present beside a large amount of small particles.[25] In this case these large particles grow at a fast rate to an even larger size, while the small matrix grains do not grow fast enough to escape the growing large grains. In principle this process can be expected to end when finally a monomodal grain size distribution of large grains is established, which grows further according to the laws of normal grain growth, see (Hillert 1965).

Therefore normal grain growth, much more than abnormal grain growth, has been a mainstay in sintering theory from its early days and still is a hot topic of research today. According to current understanding, during normal grain growth the grain size distribution adopts a time-invariant, so-called self-similar, form after a sufficiently long sintering time. That means, irrespective of the initial distribution (as long as it is monomodal and narrow enough to exclude abnormal grain growth), the final form of the distribution (attained after a sufficiently long sintering time) remains constant, i.e., the shape of the final grain size distribution (i.e., the grain size distribution after scaling, i.e., normalization with respect to a mean size) does not change any more, even though the shift towards larger grain sizes may continue. There are two questions, however: First, what is the

---

[24] The latter is sometimes called exaggerated grain growth or, going back to metallurgy, secondary recrystallization.

[25] However, abnormal grain growth also occurs when the grain boundary energies or mobilities are strongly anisotropic (Lawrence et al. 2016).

analytical form of this distribution? And second, how universal is this distribution?

A first attempt to answer the first question was made by Feltham in 1957 (Feltham 1957), who found empirically that a number of experimental distributions of grain sizes in planar sections of polycrystalline materials could be fitted well by lognormal distribution functions. The second serious attempt – and at the same time the first theoretical approach to the evolution of grain size distributions – was Hillert's aforementioned work (Hillert 1965), which was based on the LSW model approach (Lifshitz & Slyozov 1961, Wagner 1961), originally designed for second-phase particle growth in a liquid medium (assuming spherical particles in infinite dilution). Actually, Hillert was the first to apply the LSW approach to grain microstructures in dense single-phase polycrystalline materials. However, while Feltham's lognormal distribution had no theoretical foundation, Hillert's theoretically well-founded analytical model has never been observed in practice, neither experimentally (for real-world materials) nor by computer simulations (of digital microstructures), see (Streitenberger & Zöllner 2006, Zöllner & Streitenberger 2006). With respect to this situation, the search for appropriate grain size distribution functions continued and may be said to continue to date (Atkinson 1988, Jeppson et al. 2008, Streitenberger & Zöllner 2006, Zöllner & Streitenberger 2006). In particular, it should be emphasized that the Rayleigh distribution, which has been theoretically predicted by Louat in 1974 (Louat 1974), has recently been found experimentally for densely sintered magnesia-alumina spinel and yttrium-aluminum garnet ceramics (Uhlířová et al. 2016, 2018). Therefore it may be conjectured that this grain size distribution, which is identical in two-dimensional planar sections and in three-dimensional microstructures (Pabst & Uhlířová 2017), is much more universal than commonly assumed. Interestingly, Rayleigh-type grain size distributions are also frequently found in two-phase microstructures resulting from liquid phase sintering (German 1996). Unfortunately, Rayleigh distributions have been largely ignored by most authors so far, with a few notable exceptions

(Jeppsson et al. 2008, Kim 2000, Kim et al. 2003, Pande et al. 1997, Weygand et al. 1998).

A problem that has not always found due consideration in studies of grain growth kinetics is the fact that the results of two-dimensional (2D) grain growth studies have only very limited quantitative significance for three-dimensional (3D) microstructures. Therefore many of the early computer simulations of 2D systems do not have much to say about grain growth in 3D microstructures. Another, more practical, problem is that 2D grain section size distributions determined from planar sections are usually not the same as the corresponding 3D grain size distributions. It is a common misunderstanding that the latter are always shifted to larger sizes. Although it is true that sections through convex particles are usually (or more precisely, in mathematical terms, “almost always”) smaller than the maximum cross section of this particle, the probability to encounter small particles in a random section is smaller than the probability to encounter larger particles (Pabst & Uhlířová 2017). Therefore the 3D grain size distributions may be shifted to larger or smaller sizes compared to the 2D grain section size distributions, and actually there is one (and only one) distribution, for which both distributions are the same, viz. the Rayleigh distribution

$$f(x) = \frac{x}{a^2} \exp\left(-\frac{x^2}{2a^2} \right), \tag{29}$$

where x is either the 3D grain diameter or the 2D grain section diameter, and the Rayleigh parameter a determines the distribution and all its characteristic values completely, e.g., the arithmetic mean diameter $\bar{x} = a \cdot \sqrt{\pi/2}$ (Pabst & Uhlířová 2017).

In all other cases a correction is required for the so-called random section problem (Wicksell’s problem). Unfortunately, this correction is not trivial even for two-phase materials with spherical inclusions, and there is no universally acknowledged procedure to perform this correction for non-spherical contiguous polyhedral grains in single-phase polycrystalline materials. In the absence of a better alternative, it is therefore reasonable to apply those transformation matrices that have turned out to be successful in the case of spherical inclusions. It can be expected that this procedure yields realistic results in the case of isometric grains. The practical application of this procedure is not difficult and consists in determining the 2D grain section size distribution in the form of histogram (usually with 10–15 columns) and transforming the *i*-dimensional vector of 2D grain section size fractions n_i by a universal transformation matrix C_{ij} (usually Saltykov, Cruz-Orive or Woodhead) into the *j*-dimensional vector of 3D grain size fractions N_j, i.e.,

$$N_j = \sum_{i=1}^{j} n_i C_{ij} \; . \tag{30}$$

The general recipe for the construction of the transformation matrix can be found in (Pabst & Uhlířová 2017). For special cases see, e.g., (Russ & DeHoff 2000, Saltykov 1974). It has to be emphasized that without this transformation any experimental research work on 3D grain growth and its kinetics is disputable.

6. Combined-Stage Sintering Models and Master Sintering Curves

Sintering models and kinetic relations derived for the individual stages of SSS can be used in attempts to elucidate sintering mechanisms, but do not allow a unified description of the entire sintering process from start to finish. However, in 1974 Ashby (Ashby 1974) recognized the formal similarity between single-stage models of SSS and, based on this finding, he introduced the concept of so-called sintering maps, see also (Rahaman 2003). Ten years later, in 1984, DeHoff (DeHoff 1984) published the first stereology-based description of the microstructural evolution during SSS using a cell model (unfortunately based on pentagon dodecahedra, which are not space-filling), which has more recently been used to describe sintering kinetics (DeHoff 2010). Then, around 1990, several authors proposed approaches aiming towards combined-stage sintering models (Chu et al. 1991, Johnson 1989, 1990b, 1990c, Wang & Raj 1990). A major breakthrough in this field came in 1996 with the paper by Su and Johnson (Su & Johnson 1996) in which they introduced the concept of master sintering curves (MSCs), see also (Diantonio & Ewsuk 2010, Rahaman 2003). This concept is based on the combined-stage sintering model developed in Johnson's group (Hansen et al. 1992), which results in a relation between the (instantaneous) linear shrinkage on the one hand and atomistic and microstructural parameters on the other,

$$\frac{\mathrm{d}L}{L\,\mathrm{d}t} = \frac{\gamma_{SV}\Omega}{kT}\left(\frac{\Gamma_V D_V}{G^3} + \frac{\Gamma_{GB}\delta_{GB}D_{GB}}{G^4}\right), \tag{31}$$

where Γ_V and Γ_{GB} are scaling parameters that depend on the grain size, see (Hansen et al. 1992).

The MSC concept consists essentially in separating microstructural parameters from the thermal history, i.e., the temperature- and time-dependent sintering path, and results in a universal curve that describes the overall densification kinetics of compacts prepared from a selected powder by a selected green processing technique (i.e., milling, mixing, shaping, and drying). The MSC then represents the dependence of the relative density ρ_r on the (logarithm of the) thermal history, i.e., the sintering path (temperature-time schedule), expressed via the relation

$$\Psi(\rho_r) = \Theta(t, T(t)) = \int_0^t \frac{1}{T}\exp\left(-\frac{Q'}{RT}\right)\mathrm{d}t\,, \tag{32}$$

where Q' has to be interpreted as an apparent or effective activation energy, i.e., a phenomenological constant used to describe the overall sintering process, irrespective of the underlying mechanisms and individual stages. The left hand side of this relation implicitly contains all material constants and microstructural parameters, which are assumed to be unique functions of the relative density. The definition of this effective microstructure function $\Psi(\rho_r)$ given by Su and Johnson (Su & Johnson 1996) is

$$\Psi(\rho_r) = \frac{k}{D_x \gamma_{SV} \Omega} \int_{\rho_{r0}}^{\rho_r} \frac{G(\rho_r)^N}{3\rho_r \Gamma(\rho_r)} \mathrm{d}\rho_r \, , \tag{33}$$

where the diffusion coefficient D_x refers either to volume diffusion (D_V) or grain boundary diffusion (D_{GB}), the grain size exponent N is either 3 (for volume diffusion) or 4 (for grain boundary diffusion) and the scaling parameter Γ is a function of the grain size. Although not explicitly stated in (Su & Johnson 1996), it must be concluded by comparison with (Hansen et al. 1992) that in the case of volume diffusion the scaling parameter Γ in the definition of the effective microstructure function depends only on grain size, while in the case of grain boundary diffusion it depends on both grain size and grain boundary thickness. That means for grain boundary diffusion the definition of the scaling parameter Γ is different in the two papers.[26]

The MSC is usually represented by plotting the relative density (ordinate with linear scale) versus $\log\Theta(t, T(t))$. The ordinate value can be conveniently obtained via dilatometric measurements (of course with appropriate corrections in cases where the shrinkage is anisotropic), whereas the abscissa value should be calculated from a known or estimated apparent activation energy. An estimate for the latter could principally be obtained from shrinkage-rate data from either isothermal or constant-heating-rate experiments, but in practice usually the opposite is done: based on an initial estimate or qualified guess of the activation energy an

[26] It should be noted that without this redefinition the relation between macroscopic shrinkage and microscopic densification mechanisms would be dimensionally incorrect.

iterative procedure is used to bring the densification curves (i.e., relative density versus temperature) measured at different heating rates into optimal mean-square-residual coincidence (Su & Johnson 1996), so that the apparent activation energy is *de facto* extracted via the MSC approach. Once determined, the MSC can be used to predict the relative density that is expected for arbitrary sintering paths (temperature-time schedules), applied to compacts of the same material and initial microstructure.

The MSC concept has been used by many authors as a practical tool to represent experimental data obtained for compacts of a given green-processed material by different sintering paths (i.e., different temperature-time schedules). The applications range from oxide ceramic materials, such as alumina (Frueh et al. 2018, Guillon & Langer 2010, Pouchly & Maca 2010, Pouchly et al. 2013, Shao et al. 2008a, 2008b, 2009, Song et al. 2011, Su & Johnson 1996, Wang & Teng 2010), alumina-based composites with titania or zirconia (Su & Johnson 1996), barium-titanate-based ceramics (Marković & Uskoković 2009), ceria (de Florio et al. 2009), NiCuZn ferrites (Ouyang et al. 2016), lithium-sodium niobate ceramics (Mitra et al. 2013), titania (Li et al. 2007, 2008a, 2008b, Wang & Teng 2010), thoria (Kutty et al. 2004, Ray et al. 2012), thoria-urania (Banerjee et al. 2013, Kutty et al. 2003), urania and urania-SiC (Chen et al. 2014), zinc oxide (Choron et al. 2015, Ewsuk et al. 2006, Su & Johnson 1996), and zirconia (Caruso et al. 2010, Mazaheri et al. 2009, Pouchly & Maca 2010, Pouchly et al. 2013, Song et al. 2011), to non-oxide and metal materials, such as silicon carbide (Bothara et al. 2010), molybdenum (Blaine et al. 2006), nickel (Su & Johnson 1996), stainless steel (Jung et al. 2016), titanium (Robertson & Schaffer 2009, 2010), and tungsten alloys (Blaine et al. 2009, Park et al. 2006a, 2006b, 2008, 2013, Sethi et al. 2009). The MSC approach has even be applied to (super-solidus) liquid phase sintering of 316L stainless steel (Bollina et al. 2010), WC-MgO nanocomposites (Ouyang et al. 2012), liquid phase sintering of silicon nitride (Tatami et al. 2016), co-firing of NiCuZn ferrite with dielectric material based on alumina, silica and glass (Brandt et al. 2013), HfB_2-SiC composites prepared via spark plasma sintering/SPS (Enneti et al. 2012), and NiO-zirconia composite solid oxide fuel cells/SOFCs (Rayner et al.

2015). For applications of the MSC approach to spark plasma sintering/SPS see also (Guillon & Langer 2010) and for the application to electronic ceramics in general see (DiAntonio & Ewsuk 2010). On the instrumental side, a microwave dilatometer specially adapted to the construction of MSCs has been developed (Thridandapani et al. 2011). On the other hand, further theoretical developments of the MSC concept have been comparably scarce (Kiani et al. 2006, Park et al. 2009).

Despite its widespread use, the MSC concept has several weak points. One is the assumption of a single densifying sintering mechanism. This leads to the paradoxical situation that non-densifying mechanisms such as surface diffusion and evaporation-condensation are *a priori* explicitly excluded from the theoretical treatment, while in practice their presence is often *a posteriori* invoked as an argument for explaining deviations of measured data from the expected MSCs. Another point that has been criticized recently (Frueh et al. 2018), is the use of a single activation energy. However, this point is less critical than it would seem at first sight. In the first place it has to be emphasized that already the original authors called their activation energy "apparent" activation energy (Su & Johnson 1996). Secondly, it seems quite plausible to assign different activation energies to different stages of sintering, where the relative importance or even the dominance of the different mechanisms are expected to change. Nevertheless, it has to be admitted that the recent experimental findings of lower activation energies in final-stage SSS (Pouchly et al. 2013, Song et al. 2011) are somewhat surprising.

The most critical and disputable assumption of the MSC concept, however, is the assumption that the grain size is assumed to be a unique function of density. According to this assumption grain growth is always accompanied by densification or, to put it differently, grain size is not allowed to change independently of density. Although Su and Johnson (Su & Johnson 1996) have tried to underpin this assumption by experimental evidence (Gupta 1972, Wang & Raj 1990) and cite the theoretical paper by Lange and Kellett on linear arrays of spheres (Lange & Kellett 1989), as well as the paper by Coble and Gupta (Coble & Gupta 1967), who found the activation energy for grain growth and densification to be the same in intermediate-stage SSS, it is clear that at least in final-stage SSS a unique correlation between densification and grain growth cannot exist, just because even completely dense polycrystalline materials may exhibit grain growth. Of course, also in initial-stage and intermediate-stage SSS this assumption is highly questionable, which is obviously the reason why the original authors (Su & Johnson 1996) denote it as "a first approximation." A weak rationale for accepting this point of view might be the fact that the presence of pores is expected to affect grain boundary migration, but it is clear from the aforesaid that the assumption of a unique relation between (relative) density and (relative) grain size remains highly hypothetical and surely the most questionable assumption of the MSC concept. While densification and grain growth are often closely coupled or correlated, it is highly probable that different sintering paths will usually affect densification and growth in a different way. Although special cases may exist where the influence of the thermal history (i.e., the heating rate, maximum temperature and dwell time) on densification and grain growth is identical, in general it must be expected to be different. Notwithstanding all objections from the theoretical point of view, however, it has to be admitted that the MSC concept turned out to be a useful tool for sintering practice, as evidenced by its successful application to the sintering of many polycrystalline materials, both metallic and ceramic.

CONCLUSION

In this chapter the basic concepts and classical models of solid state sintering (SSS) have been summarized and critically reviewed. The history of early sintering theory has been outlined, and the highlights in this development have been emphasized, among them the works of Kuczynski (1949), Herring (1950), Kingery and Berg (1955) and Burke (1957, 1958) on sintering mechanisms and initial-stage densification kinetics, Coble's key paper (1961) on intermediate-stage and final-stage sintering and the work of Burke and Turnbull (1952) as well as Hillert (1965) on grain growth, the latter taking into account the evolution of grain size distributions in analogy to the treatment of

Ostwald ripening in two-phase systems as described by Lifshitz and Slyozov (1961) and Wagner (1961), the so-called LSW theory. Based on the modern textbook literature, mainly the authoritative treatments of German (1996) and Rahaman (2003), monographs, conference proceedings and journal papers the further development of sintering theory and practice and the current view on SSS has been briefly recalled, with a closer look on some of the key issues and hot topics of current research. In particular, it has been shown that despite enormous research efforts in grain growth theory and modeling, including massive computer simulations, the problem of normal grain growth in three-dimensional microstructures is far from being solved. Although highly sophisticated analyses in this field are now available, mainly with the papers of Streitenberger and Zöllner since 2006, which for the first time took topological constraints and their changes into due account and emphasized the important difference between simple Voronoi tessellations in virtual microstructures and spatially correlated grain microstructures in dense polycrystalline materials, we believe that an important, and maybe relatively simple, building block is still missing in grain growth theory. In particular, we are convinced that the Rayleigh distribution of grain sizes deserves much more attention in this context. Although derived in a precocious key paper by Louat (1974), the Rayleigh distribution has been largely ignored by modern authors, although recent experimental findings by Uhlířová et al. (2016, 2018) provide strong evidence for the occurrence of Rayleigh distributions of grain sizes in dense polycrystalline materials. The neglect of the Rayleigh distribution in experimental work is probably caused by the random section problem, which requires a thorough understanding of the transformation matrices used for its correction, as presented recently by Pabst and Uhlířová (2017). We think that modern research concerning the evolution of grain size distributions during SSS cannot rely on computer simulations alone, but has to compare theoretical predictions also with dense polycrystalline real-world materials, which urgently requires the application of stereology-based image analysis. Concerning densification and its kinetics, the key tool for sintering practice today is the concept of so-called master sintering curves (MSCs), which has been proposed by Su and Johnson (1996), based on the combined-stage sintering model proposed four years earlier in a paper from Johnson's group by Hansen et al. (1992). This important concept has been explained, and it has been shown that despite serious theoretical objections the MSC concept has become a successful practical tool for describing sintering processes. It has been emphasized, however, that relative density is only one of the microstructural parameters that changes during SSS and that sintering processes can also be monitored *in situ* via physical properties, such as elastic moduli measured via impulse excitation. We believe that also in this field future progress will be based on more rigorous relations between the stages of sintering and the microstructural evolution described via stereology-based image analysis.

ACKNOWLEDGMENT

This work is part of the project "Partially and fully sintered ceramics – processing, microstructure, properties, modeling and sintering theory" (GA18-17899S), supported by the Czech Science Foundation (GAČR).

REFERENCES

Aboav, D. A. *Metallography* 1970, *3,* 383-390.

Agrawal, D. Microwave sintering of ceramics, composites and metals, in *Sintering of Advanced Materials;* Fang, Z. Z. (Ed.); Woodhead Publishing: Cambridge, UK, 2010; pp 222-248.

Ashby, M. F. *Acta Metall.* 1974, *22,* 275-289.

Aste, T.; Weaire, D. *The Pursuit of Perfect Packing;* Institute of Physics: Bristol, UK, 2000, pp 20-74.

ASTM E 1876-99 *Standard test method for dynamic Young's modulus, shear modulus, and Poisson's ratio by impulse excitation of vibration;* American Society for Testing of Materials: West Conshohocken, PA, 1999.

Atkinson, H. V. *Acta Metall.* 1988, *36,* 469-491.

Banerjee, J.; Ray, A.; Kumar, A.; Banerjee, S. *J. Nucl. Mater.* 2013, *443,* 467-478.

Beck, P. A.; Kremer, J. C.; Demer, L. J.; Holzworth, M. L. *Trans. Am. Inst. Min. Metall. Engrs.* 1948, *175,* 372-376.

Belnap, J. D. Sintering of ultrahard materials, in *Sintering of Advanced Materials;* Fang, Z. Z. (Ed.); Woodhead Publishing: Cambridge, UK, 2010; pp 389-414.

Blais, C. Atmosphere sintering, in *Sintering of Advanced Materials;* Fang, Z. Z. (Ed.); Woodhead Publishing: Cambridge, UK, 2010; pp 165-188.

Blaine, D. C.; Gurosik, J. D.; Park, S. J.; Heaney, D. F.; German, R. M. *Met. Mater. Trans. A* 2006, *37,* 715-720.

Blaine, D. C.; Park, S. J.; German, R. M. *J. Am. Ceram. Soc.* 2009, *92,* 1403-1409.

Bollina, R.; Park, S. J.; German, R. M. *Powder Metall.* 2013, *53,* 20-26.

Bothara, M. G.; Atre, S. V.; Park, S. J.; German, R. M.; Sudarshan, T. S.; Radhakrishnan, R. *Met. Mater. Trans. A* 2010, *41,* 3252-3261.

Bouvard, D.; McMeeking, R. M. Estimation of the densification kinetics of particle aggregates through the simulation of the deformation of an average interparticle neck, in *Sintering Technology;* German, R. M.; Messing, G. L.; Cornwall, R. G. (Eds.); Marcel Dekker: New York, NY, 1996; pp 37-44.

Brook, R. J., Ceramic fabrication processes, in *Treatise on Materials Science and Technology Volume 9;* Wang, F. F. W. (Ed.); Academic Press: New York, NY, 1976; pp 331-364.

Brandt, B.; Naghib-Zadeh, H.; Rabe, T.; Blendell, J. *J. Am. Ceram. Soc.* 2013, *96,* 726-730.

Bross, P.; Exner, H. E. *Acta Met.* 1979, *27,* 1013-1020.

Burke, J. E. *Trans. Met. Soc. AIME* 1949, *180,* 73-91.

Burke, J. E. *J. Am. Ceram. Soc.* 1957, *40,* 80-85.

Burke, J. E. Recrystallization and sintering in ceramics, in *Ceramic Fabrication Processes;* Kingery, W. D. (Ed.); MIT Press: Cambridge, MA, 1958; pp 120-131.

Burke, J. E. The development of pore-free ceramics, in *Sintering Key Papers;* Somiya, S.; Moriyoshi, Y. (Eds.); Elsevier Applied Science: London, UK, 1990; pp 3-16.

Burke, J. E.; Turnbull, D. *Prog. Metal Phys.* 1952, *3,* 220-292.

Burke, J. E.; Rosolowski, J. H. Sintering, in *Treatise on Solid State Chemistry, Volume 4, Reactivity of Solids;* Bruce Hannay, N. (Ed.); Plenum Press: New York, NY, 1975; pp 621-660.

Caruso, R.; Mamana, N.; Benavidez, E. *J. Alloys Comp.* 2010, *495,* 570-573.

Chaston, J. C. *Platinum Metals Rev.* 1980, *24,* 70-79.

Chen, Z.; Subhash, G.; Tulenko, J. S. *J. Nucl. Mater.* 2014, *454,* 427-433.

Choron, D.; Marinel, S.; Pintault, B.; Beaudet-Savignat, S.; Macaigne, R. *J. Eur. Ceram. Soc.* 2015, *35,* 4195-4202.

Chu, M.-Y.; Rahaman, M. N.; De Jonghe, L. C.; Brook, R. J. *J. Am. Ceram. Soc.* 1991, *74,* 1217-1225.

Clark, P. W.; White, J. *Trans. Brit. Ceram. Soc.* 1950, *49,* 305-333.

Clark, P. W.; Cannon, J. H.; White, J. *Trans. Brit. Ceram. Soc.* 1953, *52,* 1-49.

Coble, R. L. *J. Appl. Phys.* 1961, *32,* 787-792.

Coble, R. L. *J. Appl. Phys.* 1961, *32,* 793-799.

Coble, R. L. *J. Appl. Phys.* 1963, *34,* 1679-1682.

Coble, R. L.; Gupta, T. K. Intermediate stage sintering, in *Sintering and Related Phenomena;* Kuczynski, G. C.; Hooten, H.; Gibson, C. F. (Eds.); Gordon & Breach Science Publishers: New York, NY, 1967; pp 423-441.

Coleman, S. C.; Beere, W. *Phil. Mag.* 1975, *31,* 1403-1413.

Crank, J. *The Mathematics of Diffusion;* Clarendon Press: Oxford, UK, 1975, pp 1-103.

De Florio, D. Z.; Esposito, V.; Traversa, E.; Muccillo, R.; Fonseca, F. C. *J. Therm. Anal. Calor.* 2009, *97,* 143-147.

DeHoff, R. T. A cell model for microstructural evolution during sintering, in *Sintering and Heterogeneous Catalysis;* Kuczynski, G. C.; Miller, A. E.; Sargent, G. A. (Eds.); Plenum Press: New York, NY, 1984; pp 23-34.

DeHoff, R. T. Path and kinetics of microstructural change in simple sintering, in *Sintering of Advanced Materials;* Fang, Z. Z. (Ed.); Woodhead Publishing: Cambridge, UK, 2010; pp 65-85.

DiAntonio, C. B.; Ewsuk, K. G. Master sintering curve and its application in sintering of electronic ceramics, in *Sintering of Advanced Materials;* Fang, Z. Z. (Ed.); Woodhead Publishing: Cambridge, UK, 2010; pp 130-161.

Dillon, S. J.; Rohrer, G. S. *Acta Mater.* 2009, *57,* 1-7.

Enneti, R. K.; Bothara, M. G.; Park, S.-J.; Atre, S. V. *Ceram. Int.* 2012, *38,* 4369-4372.

Ewsuk, K. G.; Ellerby, D. T.; DiAntonio, C. B. *J. Am. Ceram. Soc.* 2006, *89,* 2003-2009.

Exner, H. E. *Rev. Powder Met. Phys. Ceram.* 1979, *1,* 7-251.

Exner, H. E. *Powder Metall.* 1980, *4,* 203-209.

Exner, H. E. Quantitative description of microstructures by image analysis, in *Characterization of Materials Part II = Materials Science and Technology – A Comprehensive Treatment Volume 2B;* Lifshin, E. (Vol. Ed.); Cahn, R. W.; Haasen, P.; Kramer, E. J. (Series Eds.); Wiley-VCH: Weinheim, Germany, 2005; pp 281-350.

Exner, H. E.; Bross, P. *Acta Metall.* 1979, *27,* 1007-1012.

Exner, H. E.; Arzt, E. In *Physical Metallurgy;* Third edition; Cahn, R. W.; Haasen, P. (Eds.); Elsevier Science Publishers: Amsterdam, The Netherlands, 1990; pp 1885-1912.

Fang, Z. Z. (Ed.) *Sintering of Advanced Materials;* Woodhead Publishing: Cambridge, UK, 2010; pp 1-483.

Fang, Z. Z.; Wang, H. Sintering of ultrafine and nanosized particles, in *Sintering of Advanced Materials;* Fang, Z. Z. (Ed.); Woodhead Publishing: Cambridge, UK, 2010; pp 434-471.

Fayette, S.; Smith, D. S.; Smith, A.; Martin, C. *J. Eur. Ceram. Soc.* 2000, *20,* 297-302.

Feltham, P. *Acta Metall.* 1957, *5,* 97-105.

Flinn, B. D., Bordia, R. K., Rödel, J. Evolution of strength determining flaws during sintering, in *Sintering Technology;* German, R. M.; Messing, G. L.; Cornwall, R. G. (Eds.); Marcel Dekker: New York, NY, 1996; pp 13-20.

Frenkel, J. *J. Phys.* 1945, *9,* 385-391.

Frueh, T.; Ozer, I. O.; Poterala, S. F.; Lee, H.; Kupp, E. R.; Compson, C.; Atria, J.; Messing, G. L. *J. Eur. Ceram. Soc.* 2018, *xx,* xxx-xxx (in press, available online via https://doi.org/10.1016/j.jeurceramsoc.2017.12.025).

Fulrath, R. M.; Pask, J. A. (Eds.) *Ceramic Microstructures – Their Analysis, Significance, and Production;* John Wiley: New York, NY, 1968, pp 1-1008.

Geguzin, Y. E. *Wie werden aus Pulvern kompakte Werkstoffe? [How are Powders Transformed into Compact Materials?];* VEB Deutscher Verlag für Grundstoffindustrie: Leipzig, Germany, 1981, pp 1-248.

German, R. M. *Sintering Theory and Practice;* John Wiley: New York, NY, 1996, pp 1-550.

German, R. M. Thermodynamics of sintering, in *Sintering of Advanced Materials;* Fang, Z. Z. (Ed.); Woodhead Publishing: Cambridge, UK, 2010; pp 3-32.

German, R. M.; Munir, Y. A. *J. Am. Ceram. Soc.* 1976, *59,* 379-383.

Glicksman, M. E. *Phil. Mag.* 2005, *85,* 3-31.

Gouvea, D.; Smith, A.; Bonnet, J. P.; Varela, J. A. *J. Eur. Ceram. Soc.* 1998, *18,* 345-351.

Gregorová, E.; Pabst, W.; Černý, J.; Černý, M. Porous alumina and zirconia ceramics with tailored elasticity and well-defined high-temperature properties, poster at the 4th International Conference on Porous Media & Annual Meeting of the International Society for Porous Media (Interpore 2012), Purdue University, West Lafayette, IN, 14-16 May, 2012.

Guillon, O.; Langer, J. *J. Mater. Sci.* 2010, *45,* 5191-5195.

Gupta, T. K. *J. Am. Ceram. Soc.* 1972, *55,* 276-277.

Han, J. H.; Kim, D. Y. *Acta Metall. Mater.* 1995, *43,* 3185-3188.

Handwerker, C. A.; Dynys, J. M.; Cannon, R. M.; Coble, R. L. *J. Am. Ceram. Soc.* 1990, *73,* 1365-1370.

Handwerker, C. A.; Dynys, J. M.; Cannon, R. M.; Coble, R. L. *J. Am. Ceram. Soc.* 1990, *73,* 1371-1377.

Hansen, J. D.; Rusin, R. P.; Teng, M.-H.; Johnson, D. L. *J. Am. Ceram. Soc.* 1992, *75,* 1129-1135.

Heaney, D. F. Vacuum sintering, in *Sintering of Advanced Materials;* Fang, Z. Z. (Ed.); Woodhead Publishing: Cambridge, UK, 2010; pp 189-221.

Heimann, R. B. *Acta Praehist. Archaeol.* 1978, *9-10,* 79-102.

Heimann, R. B. Firing technologies and their possible assessment by modern analytical methods, in *Archaeological Ceramics;* Olin, J. S.; Franklin, A. D. (Eds.); Smithsonian Institution: Washington, DC, 1982; pp 89-96.

Heimann, R. B.; Franklin, U. M. *J. Int. Inst. Conserv. Canada* 1981, *4,* 23-45.

Herring, C. *J. Appl. Phys.* 1950, *21,* 301-303.

Herring, C. *J. Appl. Phys.* 1950, *21,* 437-445.

Hilgenfeldt, S.; Kraynik, A. M.; Koehler, S. A.; Stone H. A. *Phys. Rev. Lett.* 2001, *86,* 2685-2688.

Hillert, M. *Acta Metall.* 1965, *13,* 227-238.

Hlaváč, J. *Základy technologie silikátů [Fundamentals of Silicate Technology];* Second edition; SNTL/ALFA: Prague, Czechoslovakia, 1988, pp 13-95.

Jeppsson, J.; Agren, J.; Hillert, M. *Acta Mater.* 2008, *56,* 5188-5204.

Johnson, D. L. *J. Appl. Phys.* 1969, *40,* 192-200.

Johnson, D. L. *Phys. Sintering* 1969, *1,* B1-B22.

Johnson, D. L. Ultra-rapid sintering of ceramics, in *Science of Sintering, New Directions for Materials Processing and Microstructural Control;* Uskoković, D. P.; Palmour, H.; Spriggs, R. M. (Eds.); Plenum Press: New York, NY, 1989; pp 497-450.

Johnson, D. L. Computer simulation of the sintering of two spheres, in *Sintering Key Papers;* Somiya, S.; Moriyoshi, Y. (Eds.); Elsevier Applied Science: London, UK, 1990; pp 357-364.

Johnson, D. L. Sintering – mechanistic implications, in *Materials and Processes for Microelectronic Systems (Ceramic Transactions Volume 15);* Nair, K. M.; Pohanka, R.; Buchanan, R. C. (Eds.); American Ceramic Society: Westerville, OH, 1990; pp 77-90.

Johnson, D. L. *J. Am. Ceram. Soc.* 1990, *73,* 2576-2578.

Johnson, J. L. Sintering of refractory metals, in *Sintering of Advanced Materials;* Fang, Z. Z. (Ed.); Woodhead Publishing: Cambridge, UK, 2010; pp 356-388.

Jung, I. D.; Ha, S.; Park, S. J.; Blaine, D. C.; Bollina, R.; German, R. M. *Met. Mater. Trans. A* 2016, *47,* 5548-5556.

Kiani, S.; Pan, J.; Yeomans, J. A. *J. Am. Ceram. Soc.* 2006, *89,* 3393-3396.

Kim, B.-N. *Mater. Sci. Eng. A* 2000, *283,* 164-171.

Kim, B.-N.; Hiraga, K.; Morita, K. *Mater. Trans.* 2003, *44,* 2239-2244.

Kingery, W. D. *J. Appl. Phys.* 1960, *31,* 833-838.

Kingery, W. D.; Berg, M. *J. Appl. Phys.* 1955, *26,* 1205-1212.

Kingery, W. D.; Francois, B. The sintering of crystalline oxides, I. Interactions between grain boundaries and pores, in *Sintering and Related Phenomena;* Kuczynski, G. C.; Hooten, H.; Gibson, C. (Eds.); Gordon & Breach Science Publishers: New York, NY, 1967; pp 471-496.

Kingery, W. D.; Francois, B. The sintering of crystalline oxides, II. Densification and microstructure development in UO_2, in *Sintering and Related Phenomena;* Kuczynski, G. C.; Hooten, H.; Gibson, C. (Eds.); Gordon & Breach Science Publishers: New York, NY, 1967; pp 499-525.

Kingery, W. D.; Bowen, H. K.; Uhlmann, D. R. *Introduction to Ceramics;* Second edition; John Wiley & Sons: New York, NY, 1976; pp 448-532.

Kuczynski, G. C. *Trans. AIME* 1949, *85,* 169-178.

Kuczynski, G. C. Model experiments and the theory of sintering of two spheres, in *Sintering Key Papers;* Somiya, S.; Moriyoshi, Y. (Eds.); Elsevier Applied Science: London, UK, 1990; pp 357-364.

Kutty, T. R. G.; Khan, K. B.; Hedge, P. V.; Sengupta, A. K.; Majumdar, S.; Kamath, H. S. *Sci. Sinter.* 2003, *35,* 125-132.

Kutty, T. R. G.; Khan, K. B.; Hegde, P. V.; Banerjee, J.; Sengupta, A. K.; Majumdar, S.; Kamath, H. S. *J. Nucl. Mater.* 2004, *327,* 211-219.

Lange, F. F. De-sintering, a phenomenon concurrent with densification with powder compacts – a review, in *Sintering Technology;* German, R. M.; Messing, G. L.; Cornwall, R. G. (Eds.); Marcel Dekker: New York, NY, 1996; pp 1-12.

Lange, F. F.; Kellett, B. J. *J. Am. Ceram. Soc.* 1989, *72,* 735-741.

Lawrence, A.; Rickman, J. M.; Harmer, M. P.; Rollett, A. D. *Acta Mater.* 2016, *103,* 681-687.

Lenel, F. N. The role of plastic deformation in sintering, in *Sintering Key Papers;* Somiya, S.; Moriyoshi, Y. (Eds.); Elsevier Applied Science: London, UK, 1990; pp 543-564.

Lewis, F. T. *Am. J. Bot.* 1950, *37,* 715-721.

Li, D.; Chen, S.; Jing, Y.; Shao, W.; Zhang, Y.; Luan, W. *Sci. Sinter.* 2007, *39,* 103-110.

Li, D.; Chen, S.; Shao, W.; Ge, X.; Zhang, Y.; Zhang, S. *Mater. Lett.* 2008, *62,* 849-851.

Li, D.; Chen, S. O.; Sun, X. Q.; Shao, W. Q.; Zhang, Y. C.; Zhang, S. S. *Adv. Appl. Ceram.* 2008, *107,* 52-56.

Lifshitz, I. M.; Slyozov, V. V. *J. Phys. Chem. Solids* 1961, *19,* 35-50.

Louat, N. P. *Acta Met.* 1974, *22,* 721-724.

Louis, P.; Gokhale, A. *Metall. Mater. Trans. A* 1995, *26,* 1741-1744.

Mackenzie, J. K.; Shuttleworth, R. *Proc. Phys. Soc. B* 1949, *62,* 833-852.

Marković, S.; Uskoković, D. *J. Eur. Ceram. Soc.* 2009, *29,* 2309-2316.

Markworth, A. J. *Scripta Metall.* 1972, *6,* 957-960.

Markworth, A. J. *Met. Trans.* 1973, *4,* 2651-2656.

Marvin, J. W. *Am. J. Bot.* 1939, *26,* 280-288.

Matke, E. B. *Am. J. Bot.* 1946, *33,* 58-80.

Mazaheri, M.; Simchi, A.; Dourandish, M.; Golestani-Fard, F. *Ceram. Int.* 2009, *35,* 547-554.

Miller, K. T.; Lange, F. F. *Acta Met.* 1989, *37,* 1343-1347.

Mitra, S.; R Kulkarni, A.; Prakash, O. *Ceram. Int.* 2013, *39,* S65-S68.

Mullins, W. W. *J. Appl. Phys.* 1956, *27,* 900-904.

Mullins, W. W. *J. Appl. Phys.* 1986, *59,* 1341-1349.

Mullins, W. W. *Acta Metall.* 1989, *37,* 2979-2984.

Munir, Z. A.; Anselmi-Tamburini, U.; Ohyanagi, M. *J. Mater. Sci.* 2006, *41,* 763-767.

Munoz, N. E.; Gilliss, S. R.; Carter, C. B. *Surf. Sci.* 2004, *573,* 391-402.

Nabarro, F. R. N. Deformation of crystals by motion of single ions, in *Report of a Conference on the Strength of Solids;* Physical Society: London, UK, 1948; pp 75-90.

Niu, W.; Pan, J. Computer modelling of sintering: theory and examples, in *Sintering of Advanced Materials;* Fang, Z. Z. (Ed.); Woodhead Publishing: Cambridge, UK, 2010; pp 86-109.

Orru, R.; Licheri, R.; Locci, A. M.; Cincotti, A.; Cao, G. *Mater. Sci. Eng. R* 2009, *63,* 127-287.

Ouyang, C. X.; Zhu, S. G.; Ma, J.; Qu, H. X. *J. Alloys Comp.* 2012, *518,* 27-31.

Ouyang, C. X.; Xiao, S.; Zhu, J.; Wang, H. *Adv. Appl. Ceram.* 2016, *115,* 396-402.

Pabst, W.; Gregorová, E. *Phase Mixture Models for the Properties of Nanoceramics;* Nova Science Publishers: New York, NY, 2010; pp 1-74.

Pabst, W.; Uhlířová, T. *Ceram. Silik.* 2017, *61,* 147-157.

Pabst, W.; Gregorová, E.; Černý, M. *J. Eur. Ceram. Soc.* 2013, *33,* 3085-3093.

Pabst, W.; Gregorová, E.; Uhlířová, T. *Mater. Charact.* 2015, *105,* 1-12.

Pabst, W.; Gregorová, E.; Uhlířová, T. Processing, microstructure, properties, applications and curvature-based classification schemes of porous ceramics, in *Advances in Porous Ceramics;* Newton, A. (Ed.); Nova Science Publishers: New York, NY, 2016; pp 1-52.

Pande, C. S.; Masumura, R. A.; Marsh, S. P. *Acta Mater.* 1997, *45,* 4361-4366.

Park, S. J.; German, R. M.; Martin, J. M.; Guo, J. F.; Johnson, J. L. *Met. Mater. Trans. A* 2006, *37,* 2837-2848.

Park, S. J.; Martin, J. M.; Guo, J. F.; Johnson, J. L.; German, R. M. *Met. Mater. Trans. A* 2006, *37A,* 3337-3346.

Park, S. J.; Chung, S. H.; Martín, J. M.; Johnson, J. L.; German, R. M. *Met. Mater. Trans. A* 2008, *39,* 2941-2948.

Park, S. J.; Suri, P.; Olevsky, E.; German, R. M. *J. Am. Ceram. Soc.* 2009, *92,* 1410-1413.

Park, S. J.; Johnson, J. L.; Wu, Y.; Kwon, Y.-S.; Lee, S.; German, R. M. *Int. J. Refract. Met. Hard Mater.* 2013, *37,* 52-59.

Pask, J. A. Effect of water vapor on sintering of ceramic oxides, in *Proceedings of the First International Symposium on Hydrothermal Reactions;* Somiya, S. (Ed.); Tokyo Institute of Technology: Tokyo, Japan, 1983; pp 904-918.

Pask, J. A. Thermodynamic and geometric factors in sintering, in *Sintering Key Papers;* Somiya, S.; Moriyoshi, Y. (Eds.); Elsevier Applied Science: London, UK, 1990; pp 567-578.

Petzow, G.; Exner, H. E. *Z. Metallkunde* 1976, *67,* 611-618.

Pines, B. Ya. *Zhurnal Tekhnicheskoi Fiziki [J. Tech. Phys. USSR]* 1946, *16,* 737-743.

Poulier, C.; Smith, D. S.; Absi, J. *J. Eur. Ceram. Soc.* 2007, *27,* 475-478.

Pouchly, V.; Maca, K. *Sci. Sinter.* 2010, *42,* 25-32.

Pouchly, V.; Maca, K.; Shen, Z. *J. Eur. Ceram. Soc.* 2013, *33,* 2275-2283.

Procházka, S.; Coble, R. L. *Phys. Sinter.* 1970, *2,* 15-34.

Qian, M.; Schaffer, G. B. Sintering of aluminium and its alloys, in *Sintering of Advanced Materials;* Fang, Z. Z. (Ed.); Woodhead Publishing: Cambridge, UK, 2010; pp 291-323.

Qian, M.; Schaffer, G. B.; Bettles, C. J. Sintering of titanium and its alloys, in *Sintering of Advanced Materials;* Fang, Z. Z. (Ed.); Woodhead Publishing: Cambridge, UK, 2010; pp 324-355.

Quach, D. V.; Groza, J. R.; Zavaliangos, A.; Anselmi-Tamburini, U. Fundamentals and applications of field/current assisted sintering, in *Sintering of Advanced Materials;* Fang, Z. Z. (Ed.); Woodhead Publishing: Cambridge, UK, 2010; pp 249-274.

Rahaman, M. N. *Ceramic Processes and Sintering;* Second edition; Marcel Dekker: New York, NY, 2003; pp 425-845.

Rahaman, M. N. Kinetics and mechanisms of densification, in *Sintering of Advanced Materials;* Fang, Z. Z. (Ed.); Woodhead Publishing: Cambridge, UK, 2010; pp 33-64.

Ratke, L.; Voorhees, P. W. *Growth and Coarsening;* Springer: Berlin, Germany, 2002; pp 57-165.

Ray, A.; Banerjee, J.; Kutty, T. R. G.; Kumar, A.; Banerjee, S. *Sci. Sinter.* 2012, *44,* 147-160.

Rayner, A. J.; Clemmer, R. M. C.; Corbin, S. F.; Bordia, R. *J. Am. Ceram. Soc.* 2015, *98,* 1060-1065.

Rivier, N. *Philos. Mag. B* 1985, *52,* 795-819.

Robertson, I. M.; Schaffer, G. B. *Met. Mater. Trans. A* 2009, *40,* 1968-1979.

Robertson, I. M.; Schaffer, G. B. *Met. Mater. Trans. A* 2010, *41,* 2949-2958.

Russ, J. C.; DeHoff, R. T. *Practical Stereology;* Second edition; Kluwer Academic/Plenum Publishers: New York, NY, 2000; pp 1-109, 297-312.

Saltykov, S. A. *Stereometric Metallography;* Second edition (in Russian); State Publishing House for Metals and Sciences: Moscow, USSR, 1958; pp 1-368.

Saltykov, S. A. *Stereometrische Metallographie [Stereometric Metallography];* VEB Deutscher Verlag für Grundstoffindustrie: Leipzig, Germany; 1974; pp 1-397.

Salmang, H.; Scholze, H. *Keramik [Ceramics] – Volume I;* Sixth edition; Springer: Berlin, Germany; 1982; pp 90-91, 154-185.

Saylor, D. M.; Rohrer, G. S. *J. Am. Ceram. Soc.* 1999, *82,* 1529-1536.

Scherer, G. W. *J. Am. Ceram. Soc.* 1977, *60,* 236-239.

Scherer, G. W. *J. Am. Ceram. Soc.* 1977, *60,* 243-246.

Scherer, G. W. *J. Am. Ceram. Soc.* 1984, *67,* 709-715.

Sethi, G.; Park, S. J.; Johnson, J. L.; German, R. M. *Int. J. Refract. Met. Hard Mater.* 2009, *27,* 688-695.

Shao, W. Q.; Chen, S. O.; Li, D.; Cao, H. S.; Zhang, Y. C.; Zhang, S. S. *Sci. Sinter.* 2008, *40,* 251-261.

Shao, W. Q.; Chen, S. O.; Li, D.; Cao, H. S.; Zhang, Y. C.; Ge, X. H. *Mater. Technol.* 2008, *23,* 19-22.

Shao, W. Q.; Chen, S. O.; Li, D.; Cao, H. S.; Zhang, Y. C.; Zhang, S. S. *J. Eur. Ceram. Soc.* 2009, *29,* 201-204.

Smith, C. S. *Trans. Am. Inst. Min. Met. Eng.* 1948, *175,* 15-51.

Smith, C. S. *Metal Interfaces;* Seminar report; American Society for Metals (ASM): Cleveland, OH, 1952; p 108.

Song, X.; Lu, J.; Zhang, T.; Ma, J. *J. Am. Ceram. Soc.* 2011, *94,* 1053-1059.

Srolovitz, D. J.; Anderson, M. P.; Grest, G. S.; Sahni, P. S. *Scripta Metall.* 1983, *17,* 241-246.

Srolovitz, D. J.; Anderson, M. P.; Sahni, P. S.; Grest, G. S. *Acta Metall.* 1984, *32,* 793-802.

Stokes, R. J.; Evans, D. F. *Fundamentals of Interfacial Engineering;* Wiley-VCH: Weinheim, Germany; 1997; pp 623-669.

Streitenberger, P., Zöllner, D. *Scripta Mater.* 2015, *109,* 52-55.

Su, H.; Johnson, D. L. *J. Am. Ceram. Soc.* 1996, *79,* 3211-3217.

Suh, M. S.; Kim, M. K.; Kwon, O. J. Computer simulation of the transient grain growth with initial uniform and bimodal size distribution in two-dimensional model, in *Sintering Technology;* German, R. M.; Messing, G. L.; Cornwall, R. G. (Eds.); Marcel Dekker: New York, NY, 1996; pp 357-364.

Sutton, A. P.; Baluffi, R. W. *Interfaces in Crystalline Materials;* Clarendon Press: Oxford, UK; 1995; pp 347-621.

Swinkels, F. B.; Ashby, M. F. *Acta Met.* 1981, *29,* 259-281.

Tatami, J.; Hiratsuka, D.; Okada, S.; Komeya, K.; Wakihara, T. *J. Ceram. Soc. Jap.* 2016, *124,* 375-380.

Thridandapani, R. R.; Folz, D. C.; Clark, D. E. *Meas. Sci. Technol.* 2011, *22,* 105706 (6pp).

Thummler, F.; Thomma, W. *Met. Rev.* 1967, *12,* 69-108.

Torquato, S. *Random Heterogeneous Materials – Microstructure and Properties;* Springer: New York, NY; 2002; pp 88-95.

Torquato, S.; Truskett, T. M.; Debenedetti, P. G. *Phys. Rev. Lett.* 2000, *84,* 2064-2067.

Uhlířová, T.; Hostaša, J.; Pabst, W. *Ceram. Silik.* 2014, *58,* 173-183.

Uhlířová, T.; Pabst, W.; Gregorová, E.; Hostaša, J. *J. Eur. Ceram. Soc.* 2016, *36,* 2319-2328.

Uhlířová, T.; Nečina, V.; Pabst, W.; Diblíková, P. Microstructure and grain size distributions in magnesia-alumina spinel ceramics prepared by spark plasma sintering, in *Proceedings of the 12th Pacific Rim Conference on Ceramic and Glass Technology;* Ceramic Transactions, Volume 264; Singh, D.; Fukushima, M.; Kim, Y. W. (Eds.); American Ceramic Society/John Wiley & Sons: Hoboken, NJ, 2018; pp 41-52.

Usković, D.; Exner, H. E. *Sci. Sinter.* 1977, *9,* 265-303.

Vasilos, T.; Rhodes, W. Solids processing of fine grain size ceramics, in *Characterization of Ceramic Materials;* Saito, S.; Kotera, Y.; Somiya, S. (Eds.); First US-Japan Seminar on Basic Science of Ceramics: Tokyo, Japan, 1969; pp 93-140; reprinted in *Sintering Key Papers;* Somiya, S.; Moriyoshi, Y. (Eds.); Elsevier Applied Science: London, UK, 1990; pp 741-773.

Von Neumann, J. Discussion of Smith, C. S., Grain shape and other metallurgical aspects of topology, in *Metal Interfaces;* Seminar report; American Society for Metals (ASM): Cleveland, OH, 1952; pp 65-110.

Wagner, C. *Z. Elektrochem.* 1961, *65,* 581-591.

Wang, J.; Raj, R. *J. Am. Ceram. Soc.* 1990, *73,* 1172-1175.

Wang, S. Y.; Teng, M. H. *J. Alloys Comp.* 2010, *504,* S336-S339.

Wellner, P.; Exner, H. E. *Phys. Sinter.* 1973, *5,* 25-37.

West, J.; Sears, J. W.; Smith, S.; Carter, M. Photonic sintering – an example: photonic curing of silver nanoparticles, in *Sintering of Advanced Materials;* Fang, Z. Z. (Ed.); Woodhead Publishing: Cambridge, UK, 2010; pp 275-288.

Weygand, D.; Brechet, Y.; Lepinoux, J. *Philos. Mag. B* 1998, *78,* 329-352.

Whittemore, O. J.; Varela, J. A. Pore growth and shrinkage during sintering, in *Sintering Key Papers;* Somiya, S.; Moriyoshi, Y. (Eds.); Elsevier Applied Science: London, UK, 1990; pp 777-792.

Zöllner, D. *Comp. Mater. Sci.* 2013, *79,* 759-762.

Zöllner, D.; Streitenberger, P. *Scipta Mater.* 2006, *54,* 1697-1702.

Zöllner, D.; Rios, P. R. *Acta Mater.* 2017, *130,* 147-154.

In: Polycrystalline Materials
Editor: Jim Olson
ISBN: 978-1-53613-864-1

Chapter 2

MODELLING AND ANALYSIS OF MORPHOLOGICAL ANISOTROPY OF POLYCRYSTALLINE MATERIALS

Maroš Martinkovič[1,*] and Stanislav Minárik[2,†]
[1]Institute of Production Technologies,
Slovak University of Technology in Bratislava,
Faculty of Materials Science and Technology in Trnava,
Trnava, Slovak Republic
[2]Advanced Technologies Research Institute,
Slovak University of Technology in Bratislava,
Faculty of Materials Science and Technology in Trnava,
Trnava, Slovak Republic

ABSTRACT

Grain boundary - i.e., the interface between individual grains is one of the principal microstructural parameters in the polycrystalline

---

* Corresponding Author Email: maros.martinkovic@stuba.sk.
† Corresponding Author Email: stanislav. minarik@stuba.sk.

materials. In case of isotropic structure, the grains have isometric dimensions and mean grain size or grain size distribution and specific surface area of grain boundaries are sufficient structural characteristics. In case of anisotropic structure the grains have anisometric dimensions caused, for instance, by plastic deformation and it is necessary to describe their orientation. In this context the first question is how can the spatial orientation of grains themselves be defined generally and exactly, the second question is how can the strain be estimated from this orientation. It is possible to suspect intuitively that grain surface orientation in this sense can be evaluated by the prevailing orientation of grain surface area elements. The anisotropic microstructure can be decomposed into isotropic, planar and/or linear oriented components - specific surface area of grain boundaries and these parameters are experimentally measured using stereology. But real state of grain shape is quite impossible to describe, therefore model of grain shape based on an idealized shape (tetrakaidecahedron) of grains has been proposed. One can easily understand that there is a correlation between the change in the grain orientation and grain deformation and this correlation was mathematically demonstrated. Finally, we interpret the quantity of grain orientation corresponding with the experimental measured value, the analysis of local deformation could be performed experimentally on the basis of our results.

Keywords: polycrystalline structure, grain shape, grain boundary, orientation, deformation, stereology

1. INTRODUCTION

Mechanical working leads to final properties of forming pieces, which are affected by conditions of production technology. Recently the forming is in general based on macroscopic effects of deformation, but these are not corresponding fully with microscopic structural changes. The macroscopic strain does not reflect microstructural changes in volume of deformed parts of material especially when only surface layers are deformed. To obtain exact knowledge about the material's structural change in the volume it is necessary to use one of microscopic or macroscopic methods, for instance methods based on acoustic emission, microhardness measure, dislocation analysis, slip band observation, micromeshes method or macroscopic screw method, but none of them give analytical expressions relating structural parameters and strain in each location of material bulk. The development of analytical formulas relating to structural parameters and strain in each position inside the bulk sample is essential for a quantitative description of local changes induced by microstructure plastic deformation. Grain boundary - i.e., the interface between individual grains is one of the principal microstructural parameters in the polycrystalline material (metal, alloy). In case of isotropic structure, the grains have isometric dimensions and mean grain size or grain size distribution and specific surface area of grain boundaries are sufficient structural characteristics. In case of anisotropic structure the grains have anisometric dimensions caused by plastic deformation and it is necessary to describe their orientation. In this context the first question is how the spatial orientation of grains themselves can be defined generally and exactly, the second question is how the strain can be estimated from this orientation. It is possible to suspect intuitively that grain surface orientation in this sense can be evaluated by the prevailing orientation of grain surface area elements.

Experimental observation and quantitative evaluation of shape and orientation of grains, domains or cells in grained, foam and cellular structures have already been at the center of several author's attention for some time. Analysis of experimental data shows that grain boundaries play an important role in the process of deformation [1-4]. Various experimental procedures to characterize size, deformation, orientation and position of single grains within the bulk of a polycrystalline sample have been developed [5-9]. Most of these methods are currently considered to be standard experimental techniques. Calculations of grain boundary distortion during plastic straining have proven to be important in the theory of crystal plasticity and stereology [10-14]. Modelling of single crystal plasticity by numerical methods allows discussion of deviations of experimental results from ideal crystal behavior [15-17]. Calculation of change in the grain surface area edge per unit volume as a function of common deformations is important in the production of steels [18]. Amount of grain surface (S_V) and grain edge (L_V) per unit volume are presented as parameters important in kinetic theory since both of them are heterogeneous nucleation sites.

It is clear from previous work that there are a few ways to evaluate the grain orientation experimentally. However in practice the most suitable way to determine the degree of grain orientation is the scalar measurement of anisotropy based on stereological principle. Anisotropic microstructure consists of (planar or linearly) oriented components which can be easily evaluated using stereological methods [19]. In addition we suggest that results of such measurement, i.e., degree of grain orientation in volume of plastically deformed structure can be converted to values of local deformation parameters. In that case it is sufficient to measure the degree of grain orientation to obtain a value of local strain. It is completely impossible to describe actual shape of the grain in material structure exactly. Therefore deformation of various idealized grain shapes can be investigated [20]. For instance deformation of spheres can be considered since spheres are not space filling and have no edges. Deformation of cubes can also be investigated and mathematical analysis can be simplified, but cubes are clearly poor approximations of the shapes of real grains, but the shape of a tetrakaidecahedron which is space filling can be regarded as approximately identical with the shape of the grains observed metallographically in the undeformed state. One of the methods using the idealized tetrakaidehadreonal shape of a grain is based on a dependence of the ratio of the specific surface area of grain boundaries in the deformed state S_V and the undeformed state S_{V0} (respectively specific length of edge L_V and L_{V0}) to strain [21-22]. The method requires knowledge of the parameter of structure in the case of zero value of initial deformation, which is unknown in most cases and the value of this parameter is not the same in the whole volume of pieces and it depends on grain size. Our presented method can also be used for idealized tetrakaidecahedronal grain shape, but in this case orientation of the undeformed grain was not zero. This orientation depends on the direction of the grain deformation [23]. So we tried to apply the mean value of all directions. Investigation of this method is very interesting, because it has been shown that such a deformation mechanism, i.e., elongation of grains under applied load, exists in a real-world material, as it can be seen in the cutting zone during turning [24] (Figure 1) or in an arbitrary place of forming pieces [23] (drawing tube in Figure 11).

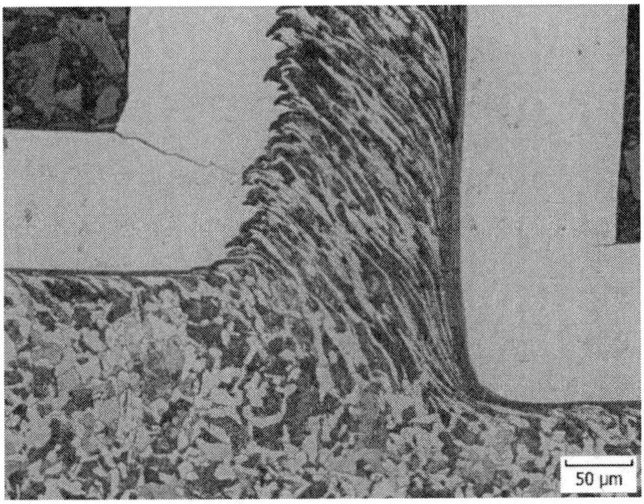

Figure 1. Local plastic deformation observable in the cutting zone of turned steel workpiece.

2. Measuring Methods

2.1. Definition of Deformation

There are only three principal schemes of elementary bulk deformation, which allows very good evaluation of the strain analysed in place of the forming body (see Figure 2). They are basic indicators for analysis and evaluation of the deformation state caused by external load.

The scheme of deformation presented in Figure 2a relates for instance to tube drawing, $\varphi_1 > 0$, φ_2, $\varphi_3 < 0$ and $\varphi_1 = - \varphi_2 - \varphi_3$, in the case when $\varphi_3 = \varphi_2$ (for instance wire drawing) $\varphi_2 = \varphi_3 = -\varphi_1/2$. The scheme of deformation presented in Figure 2b relates for instance to combination drawing and rolling, $\varphi_1 < 0$, φ_2, $\varphi_3 > 0$ and $\varphi_1 = - \varphi_2 - \varphi_3$, in the case when $\varphi_3 = \varphi_2$ (for instance upsetting, rolling) $\varphi_2 = \varphi_3 = -\varphi_1/2$. The third scheme of deformation (presented in Figure 1c when $\varphi_3 = 0$) relates for instance to bending, $\varphi_1 > 0$, $\varphi_2 < 0$ and $\varphi_1 = -\varphi_2$. True (logarithmic) strain φ is defined from its dimension before deformation l_0 and after deformation l and from relative strain δ as:

$$\varphi = \int_{l_0}^{l} \frac{1}{l} dl \qquad \varphi = \ln(1+\delta) \qquad \delta = \int_{l_0}^{l} \frac{1}{l_0} dl \tag{1}$$

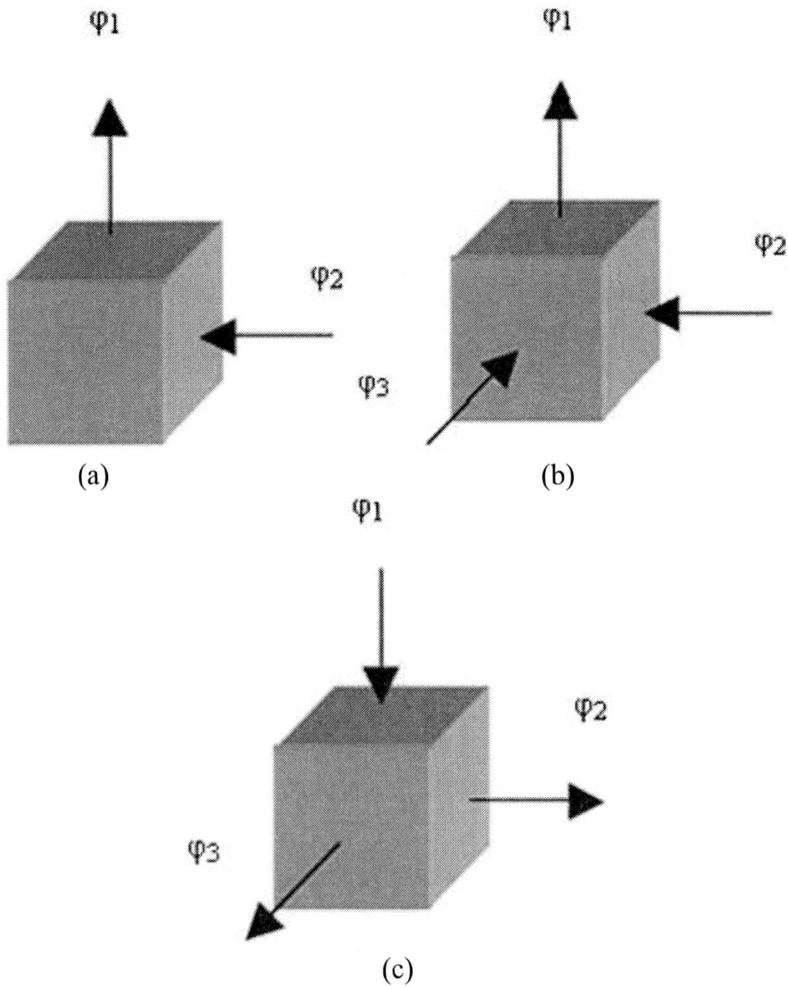

Figure 2. Principal schemes of elementary bulk deformation.

2.2. Orientation Measurement

Direction of the grain boundary's orientation is the same as the direction of deformation. If the deformation scheme is known, grain boundaries can be decomposed into isotropic, planar and linear oriented components. Saltykov stereology methods with oriented test lines were used [19]. In the case of linear deformation, for instance wire drawing (Figure 3a – direction of deformation is marked by arrows), grain boundary orientation can be observed in a plane parallel to drawing direction (according to deformation scheme Figure 2a parallel to φ_1) - linear orientation, in the case of rolling or upsetting (Figure 3b) in a plane perpendicular to the deformation plane (according to deformation scheme Figure 2b parallel to φ_1) - planar orientation. On a metallographic cut, test lines are placed perpendicular and parallel to the grain boundary orientation direction affected by the straining. From the specific number (number to unit of length) of parallel test lines (1 in Figure 3) intersections with grain boundaries $(P_L)_P$ and the perpendicular lines (2 in Figure 3) $(P_L)_O$ was total specific surface area (area to unit test volume) $(S_V)_{TOT}$ of grains estimated - according to equation (2) in the case of planar orientation and according to equation (4) in the case of linear orientation. The planar oriented part of the specific surface area $(S_V)_{OR}$ of grains was estimated according to equation (3), the linear oriented part according to equation (5). Degree of grain boundaries orientation O was estimated as $(S_V)_{OR}$ to $(S_V)_{TOT}$ ratio.

$$(S_V)_{TOT} = (P_L)_O + (P_L)_P \tag{2}$$

$$(S_V)_{OR} = (P_L)_O - (P_L)_P \tag{3}$$

$$(S_V)_{TOT} = \frac{\pi}{2}(P_L)_O + \left(2 - \frac{\pi}{2}\right)(P_L)_P \tag{4}$$

$$(S_V)_{OR} = \frac{\pi}{2}[(P_L)_O - (P_L)_P] \tag{5}$$

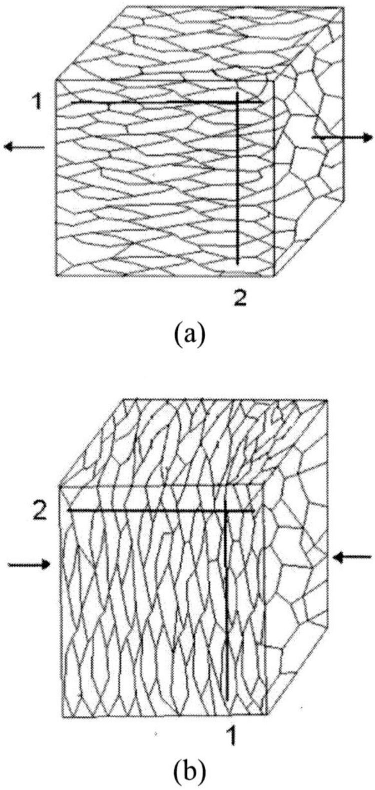

Figure 3. Anisotropic structure due to various types of deformation- linear a), planar b).

2.3. Orientation – Deformation Conversion Model

Our conversion method is based on the analysis of orientation – deformation relationship of a grain. Dependence of true strain φ on the value of orientation o was derived from three basic equations – definition of deformation (1), definition of degree of orientation (7) and invariability of volume ($V_0 = V$ - initial volume is equal to volume after plastic deformation, $\varphi_1 + \varphi_2 + \varphi_3 = 0$). The solution of the system includes one free parameter – grain size. Solution of the system of equations for used idealised grain shapes is independent of the initial dimension of the grain – strain depends only on the shape of grain and it does not depend on its dimension. As a result the method enables estimation of local plastic deformation from the estimation of microstructure anisotropy in an arbitrary place on the body with an arbitrary state of initial deformation.

2.4. Evaluation of Local Plastic Deformation in the Tetrakaidecahedronal Model of Grained Structure from Stereological Principles

Grained structure is of paramount importance to materials science. It provides an important example of a space-filling irregular network structure that also occurs in foams as well as in certain biological tissues. Therefore, the searching for characteristics of individual grains is of fundamental importance. It is absolutely impossible to describe the actual shape of each grain in a grained structure analytically and for this reason approximations based on various idealized grain shapes have been investigated. However approximate methods usually assume that all the grains are exactly identical in shape and size.

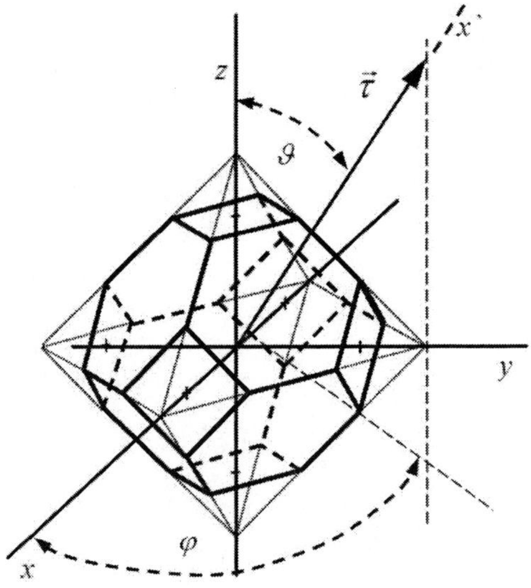

Figure 4. Undeformed tetrakaidecahedron. In case of elongation we consider vector $\vec{\tau}$ oriented parallel to the x` axis.

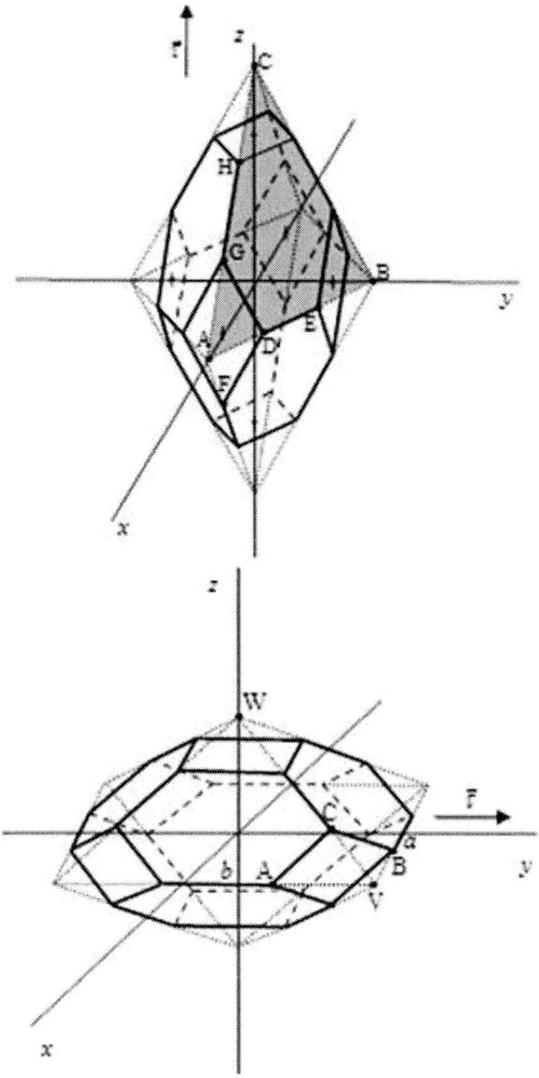

Figure 5. Deformed tetrakaidecahedron. Configurations for a) $\vec{\tau} = [0,0,1]$, b) $\vec{\tau} = [0,1,0]$.

We now consider a single tetrakaidecahedronal grain which has 24 vertices, 8 hexagonal and 6 square faces, and 36 edges, each of length *a*. The considered grain is shown in the Figure 4, where standard notation for coordinates is used i.e., $X_1 = x$, $X_2 = y$ and $X_3 = z$. Surface area (S) and volume (V) of the undeformed tetrakaidecahedron can be calculated as:

$$S = \left(6 + 12\sqrt{3}\right)a^2, V = 8\sqrt{2}a^3. \quad (6)$$

During an applied mechanical loading in different orientations and the influences from neighboring grains most of the grains are subjected to a deformation. The deformation of grain results in a change in orientation and amount of grain surface area which leads to local heterogeneous plasticity. This is evident in the case of the typical grain prolongation in the direction of the loading force is visible (see Figure 5). Let's define the degree of the whole grain-surface orientation in relation to the vector $\vec{\tau}$ as [23]:

$$o = 1 - 2\frac{1}{S}\int_{(Surface)} \left|\vec{\tau} \cdot d\vec{S}\right|. \quad (7)$$

We believe that variable o defined by (7) enables to analyze grain-surface area distortion induced by deformation. Moreover the fundamental assumption used to establish the theory of plasticity is that plastic deformation is volume preserving. Next we calculate degree of deformed grain-surface orientation O using formula (7) in relation to the vector of the applied deforming force in the case of plastic grain deformation in which the change in volume (V) of the grain is zero (i.e., $V_{undefor} = V_{defor}$).

Once the grain shape is defined, it is possible to homogeneously and plastically deform it by applying an appropriate mathematical transformation. Consider the uniform deformation of a tetrakaidecahedron in the direction determined by unit vector $\vec{\tau}$. This vector $\vec{\tau}$ is determined by the direction of the loading force. Generally the elongation ratio ε is defined as the ratio of the stretched length to the unstretched length of the tetrakaidecahedron. The relative increase in length L of the line of original length unity determines elongation δ (first consider $\delta > 1$):

$$\delta = \frac{\Delta L}{L} = \frac{L - L_0}{L_0} \tag{8}$$

The change of total surface of the tetrakaidecahedron at plastic deformation and the relationship between the degree of orientation O and deformation parameter δ can be investigated by means of procedure described in the previous text.

Position vectors of each point on the surface of grain in undeformed configurations are denoted as:

$$\vec{R}(u,v) = \left[X_1(u,v), X_2(u,v), X_3(u,v)\right] \text{ , } u \in \langle u_1, u_2 \rangle \;\; v \in \langle v_1, v_2 \rangle \tag{9}$$

Let the position vectors of each point on the surface of the same grain in deformed configurations are:

$$\vec{r}(u,v) = \left[x_1(u,v), x_2(u,v), x_3(u,v)\right] \text{ , } u \in \langle u_1', u_2' \rangle \;\; v \in \langle v_1', v_2' \rangle \tag{10}$$

If we substitute vector $\vec{\tau} = \tau_x \vec{i} + \tau_y \vec{j} + \tau_z \vec{k}$ into the formula (7) and apply (9) and (10) the degree of grain-surface orientation o can be written in the next coordinate form:

$$o = 1 - \frac{2\int\limits_{v_1}^{v_2}\int\limits_{u_1}^{u_2}\left|\left(\frac{\partial X_2'}{\partial u}\frac{\partial X_3'}{\partial v} - \frac{\partial X_2'}{\partial v}\frac{\partial X_3'}{\partial u}\right)\tau_x - \sqrt{(\delta+1)^3}\left(\frac{\partial X_1'}{\partial u}\frac{\partial X_3'}{\partial v} - \frac{\partial X_1'}{\partial v}\frac{\partial X_3'}{\partial u}\right)\tau_y + \sqrt{(\delta+1)^3}\left(\frac{\partial X_1'}{\partial u}\frac{\partial X_2'}{\partial v} - \frac{\partial X_1'}{\partial v}\frac{\partial X_2'}{\partial u}\right)\tau_z\right| du dv}{\int\limits_{v_1}^{v_2}\int\limits_{u_1}^{u_2}\sqrt{\left(\frac{\partial X_2'}{\partial u}\frac{\partial X_3'}{\partial v} - \frac{\partial X_3'}{\partial u}\frac{\partial X_2'}{\partial v}\right)^2 + (\delta+1)^3\left\{\left(\frac{\partial X_1'}{\partial u}\frac{\partial X_3'}{\partial v} - \frac{\partial X_3'}{\partial u}\frac{\partial X_1'}{\partial v}\right)^2 + \left(\frac{\partial X_1'}{\partial u}\frac{\partial X_2'}{\partial v} - \frac{\partial X_2'}{\partial u}\frac{\partial X_1'}{\partial v}\right)^2\right\}} du dv}, \tag{11}$$

where X_1, X_2, X_3 are coordinates of each point on the surface of the grain in undeformed configurations. Parametric representation of individual segments on the tetrakaidecahedron surface in the given coordinate frame shown in Figure 4 can be written in the next simplified form. Parametric representation of hexagons:

$$\vec{R}=\left[x_1(u,v),x_2(u,v),x_3(u,v)\right]=\left[(-1)^n u,(-1)^m v,(-1)^{q+1}\frac{3}{\sqrt{2}}a-(-1)^q(u+v)\right]$$

where:

$$v\in\left\langle\frac{1+(-1)^p}{2\sqrt{2}}a;\frac{2}{\sqrt{2}}a\right\rangle,$$

$$u\in\left\langle\frac{1+(-1)^{p+1}}{2}\left(\frac{a}{\sqrt{2}}-v\right);\frac{1+(-1)^{p+1}}{\sqrt{2}}a+\frac{1+(-1)^p}{2}\left(\frac{3a}{\sqrt{2}}-v\right)\right\rangle \tag{12}$$

and n, m, p, $q = 1, 2$. Parametric representation of squares:

$$\vec{R}=\left[(-1)^w u;(-1)^s v;(-1)^c\frac{2}{\sqrt{2}}a\right],\quad \vec{R}=\left[(-1)^w\frac{2}{\sqrt{2}}a;(-1)^s v;(-1)^c u\right],$$

$$\vec{R}=\left[(-1)^w u;(-1)^s\frac{2}{\sqrt{2}}a;(-1)^c v\right]$$

where: $v\in\left\langle 0;\frac{a}{\sqrt{2}}\right\rangle$, $u\in\left\langle 0;\frac{a}{\sqrt{2}}-v\right\rangle$

and w, s, $c = 1, 2$. (13)

Coordinates of vectors (12) and (13) must be applied in (11) when the degree of orientation O of deformed tetrakaidecahedron is evaluated. Vector $\vec{\tau}$ parallel with the x-axis was applied in the formula (11), i.e., $\vec{\tau}=[1,0,0]$. However, we are often interested in the value of grain orientation o with respect to any vector $\vec{\tau}$. Any deformation can be easily taken into account in (11) by a rotational transformation of coordinates by in such a way that the x-axis of the new coordinate system to be oriented in the direction of vector $\vec{\tau}$:

$$X_i' = C_{ij} X_j \quad , \tag{14}$$

where $C_{ij} = C_{ij}(\varphi, \vartheta)$ are coefficients of the ordinary rotational matrix. Calculation of the degree of orientation is relatively simple because coordinates of the position vectors X_i for each point on the tetracaidecahedron surface are linear functions of parameters u, v (there are only planar areas on the tetracaidecahedron surface). Therefore derivatives of coordinates equals constants, i.e., integrated square roots in (11) are independent of the integration variables u, v and they can be excluded before the integral.

It is necessary to integrate through each flat area on the tetracaidecahedron surface separately. Therefore we can easily transform coordinates of position vectors (12) and (13) by means of (14) and then substitute them to formula (11). When we apply this procedure on each flat segments in the surface of the tetrakaidekahedron separately, an applicable formula is obtained. Deformations of tetrakaidecahedra in a variety of orientations using the principles of stereology can be investigated by means of this formula. Various configurations can be considered by applying different a combination of φ and ϑ (see Figure 4) and deformation of grains in many directions can be investigated. As an example the relationship between orientation and deformation in different cases was calculated. Examples of some obtained results are shown graphically in Figure 6.

Any randomly oriented tetrakaidecahedronal grain in the model of a bulk grained structure can be included into the evaluation of o if the corresponding vector $\vec{\tau}$ will be applied in the calculation presented above. The model of a grained structure is considered to be a large number of randomly oriented tetrakaidecahedrons and therefore all values of φ and ϑ must be considered if the orientation of the whole structure is evaluated. We suppose that the average value of degree of a single grain orientation is acceptable for the evaluation of the degree of the whole grained structure orientation.

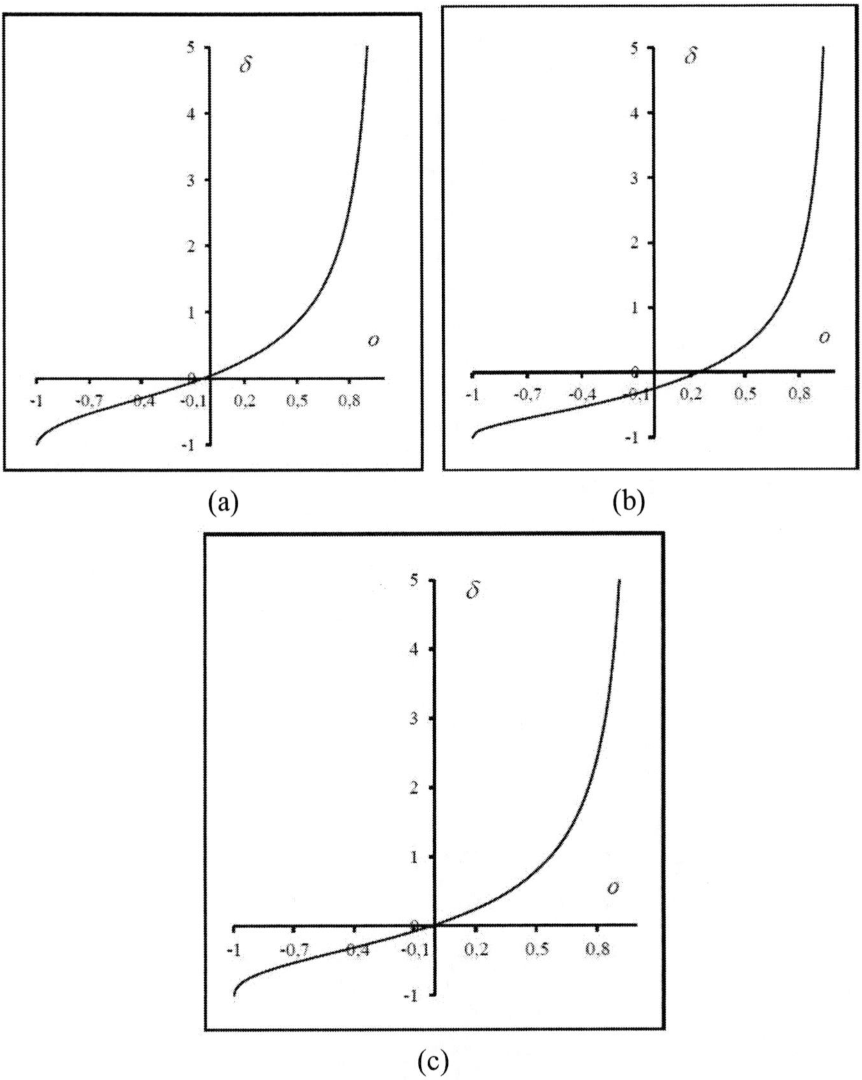

Figure 6. Elongation (contraction) δ of single tetrakaidecahedron as a function of degree of orientation o calculated in relation to some chosen vectors $\vec{\tau}$. Graphs were calculated using formula (11) for vectors $\vec{\tau}$ lying in coordinate plane xy, i.e., $\vartheta = 0^{o}$ while: a) $\varphi = 0^{o}$, b) $\varphi = 40^{o}$, c) $\varphi = 80^{o}$.

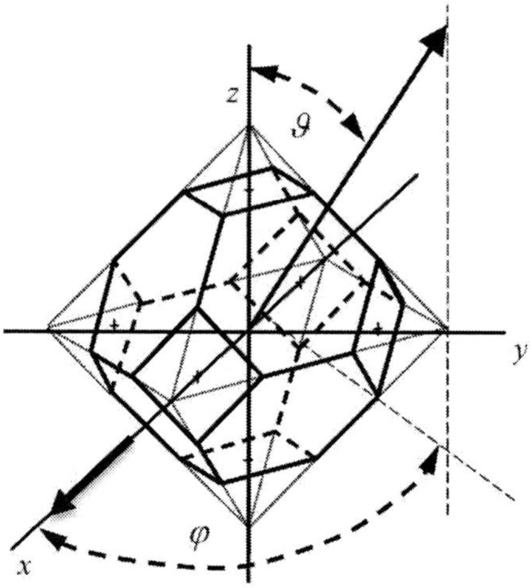

Figure 7. Undeformed tetrakaidecahedron. In case of elongation (contraction) we consider vector $\vec{\tau}$ oriented (parallel) perpendicular to the x axis.

Calculation of the average value of the grain orientation in the case of tetrakaidecahedron elongation can be realized using the scheme shown in the Figure 7. If we determine the average value o_{avr} in the plane xy in configuration when the angle ϑ changes in the range from 0 to 2π.

$$o_{avr}\left(\delta,\varphi,\vartheta,\theta\right)=\frac{1}{2\pi}\int_{0}^{2\pi} o\left(\delta,\varphi,\vartheta,\theta\right)d\theta \tag{15}$$

Different values of angle ϑ represent a different orientation of tetrakaidecahedron in the model of a grained structure. Applying formulas (11) and (14) in the equation (15) we get:

$$o_{avr}\left(\delta,\varphi,\vartheta\right)=1-\frac{\Pi\left(\varphi,\vartheta,\theta\right)+\frac{3}{2}\sum_{m=1}^{2}\sum_{n=1}^{2}\sum_{q=1}^{2}K_{m,n,q}\left(\varphi,\vartheta,\theta\right)}{\eta\left(\delta,\varphi,\vartheta\right)+\chi\left(\delta,\varphi,\vartheta\right)+\lambda\left(\delta,\varphi,\vartheta\right)+\frac{3}{2}\sum_{m=1}^{2}\sum_{n=1}^{2}\sum_{q=1}^{2}\sqrt{\beta_{m,n,q}\left(\varphi,\vartheta\right)+v_{m,n,q}\left(\delta,\varphi,\vartheta\right)}} \tag{16}$$

where:

$$\Pi(\varphi,\vartheta,\theta)=4\left\{\left|sin(\varphi)sin(\theta)\right|+\left|sin(\varphi)cos(\theta)-cos(\varphi)cos(\vartheta)sin(\theta)\right|+\left|sin(\varphi)cos(\vartheta)sin(\theta)-cos(\varphi)cos(\theta)\right|\right\}$$

$$K_{m,n,q}(\varphi,\vartheta,\theta)=3\left|(-1)^{q}\left\{(-1)^{n}cos(\varphi)-(-1)^{m}sin(\varphi)\right\}cos(\theta)-\left\{(-1)^{m+n}sin(\vartheta)-(-1)^{m+q}cos(\varphi)cos(\vartheta)-(-1)^{n+q}sin(\varphi)cos(\vartheta)\right\}sin(\theta)\right|$$

$$\eta(\delta,\varphi,\vartheta)=2\sqrt{cos^{2}(\vartheta)+(\delta+1)^{3}sin^{2}(\vartheta)}$$

$$\chi(\delta,\varphi,\vartheta)=2\sqrt{cos^{2}(\varphi)sin^{2}(\vartheta)+(\delta+1)^{3}\left[sin^{2}(\varphi)+cos^{2}(\varphi)cos^{2}(\vartheta)\right]}$$

$$\lambda(\delta,\varphi,\vartheta)=2\sqrt{sin^{2}(\varphi)sin^{2}(\vartheta)+(\delta+1)^{3}\left[sin^{2}(\varphi)cos^{2}(\vartheta)+cos^{2}(\varphi)\right]}$$

$$\beta(\delta,\varphi,\vartheta)=1+(-1)^{q}\left[(-1)^{m}sin(\varphi)+(-1)^{n}cos(\varphi)\right]sin(2\vartheta)+(-1)^{m+n}sin(2\varphi)sin^{2}(\vartheta)$$

$$\nu_{n,n,q}=(\delta+1)^{3}\left\{2-(-1)^{q}\left[(-1)^{n}cos(\varphi)+(-1)^{m}sin(\varphi)\right]sin(2\vartheta)+(-1)^{m+n}sin(2\varphi)cos^{2}(\vartheta)-(-1)^{m+n}sin(2\varphi)\right\}$$

We expect deformation in the direction of the bold arrow (see Figure 7) and therefore φ = 0o and ϑ = 90o. This corresponds with the concept of linear orientation. When we put these values into the equation (16) we get:

$$o_{avr}(\delta)=\frac{6\sqrt{2(\delta+1)^{3}+1}+2\sqrt{\delta+1}(\delta+1)-13}{6\sqrt{2(\delta+1)^{3}+1}+2\sqrt{\delta+1}(\delta+1)+1} \quad (17)$$

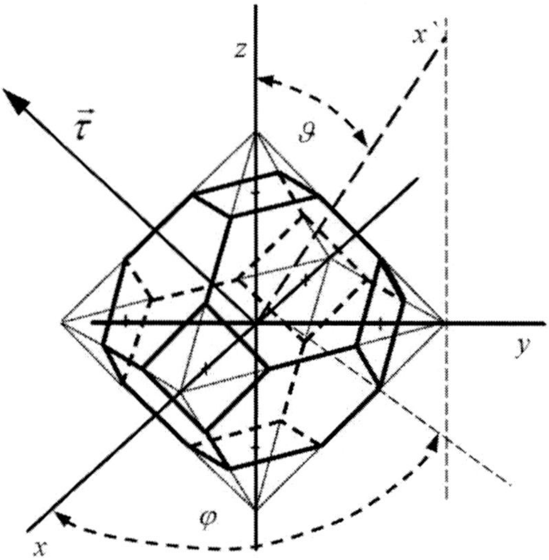

Figure 8. Undeformed tetrakaidecahedron. In case of contraction we consider vector $\vec{\tau}$ oriented perpendicularly to the x` axis.

In the previous text we evaluated the orientation of the grained structure using the formula (7) in a case where an extension of this structure caused by the deformation force is observed, i.e., if $0 < \delta$. This corresponds with the concept of linear orientation in Saltykov stereology method. In the case when there is a contraction of the structure under the deformation force, i.e., $-1 < \delta < 0$, we consider it appropriate to evaluate orientation o in relation to any vector $\vec{\tau}$ oriented perpendicular to the deformation force, i.e., $\vec{\tau} = [0, cos(\theta), sin(\theta)]$ where θ is the angle between the axis y and vector $\vec{\tau}$. This angle changes from 0 to 2π (see Figure 8) and we expect deformation in the opposite direction to the bold arrow (see Figure 7). In that case applying formulas (11) and (14) in the equation (15) we get:

$$o_{avr}(\delta,\varphi,\vartheta,\theta)=\frac{1}{2\pi}\int_0^{2\pi}\left\{1-\frac{(\delta+1)^{\frac{3}{2}}\left\{\Pi(\varphi,\vartheta,\theta)+\frac{3}{2}\sum_{m=1}^{2}\sum_{n=1}^{2}\sum_{q=1}^{2}K_{m,n,q}(\varphi,\vartheta,\theta)\right\}}{\eta(\delta,\varphi,\vartheta)+\chi(\delta,\varphi,\vartheta)+\lambda(\delta,\varphi,\vartheta)+\frac{3}{2}\sum_{m=1}^{2}\sum_{n=1}^{2}\sum_{q=1}^{2}\sqrt{\beta_{m,n,q}(\varphi,\vartheta)+\nu_{m,n,q}(\delta,\varphi,\vartheta)}}\right\}d\theta \quad (18)$$

and after putting angles:

$$o_{avr}(\delta)=\frac{1+6\sqrt{1+2(\delta+1)^3}-2\left(\frac{4-\pi+12\sqrt{2}}{\pi}\right)\sqrt{(\delta+1)^3}}{1+6\sqrt{1+2(\delta+1)^3}+2\sqrt{(\delta+1)^3}} \tag{19}$$

This corresponds with the concept of planar orientation. However the principle of calculation is exactly the same as in the previous case.

Any randomly oriented tetrakaidecahedronal grain in the model of the bulk grained structure can be included into the evaluation of the whole structure orientation o_s if the corresponding (randomly oriented) vector $\vec{\tau}$ will be applied in the calculation presented above. The model of the grained structure is considered to be a large number of randomly oriented tetrakaidecahedrons and therefore all values of φ and ϑ must be considered if the orientation of the whole structure is to be evaluated. We suppose that the average value of the degree of the grained structure linear orientation will be calculated as:

$$o_s(\delta)=\frac{1}{2\pi^2}\int_0^{\pi}\int_0^{2\pi} o(\varphi,\vartheta,\delta)\,d\varphi\, d\vartheta \tag{20}$$

This is acceptable for the evaluation of the degree of the whole grained structure orientation. However the calculation cannot be performed analytically in this general case and numerical methods must be applied.

Models of grained structures based on the specific idealized shape of grains allow us to examine the relationship between grain orientation and local deformation but the problem of non-zero grain orientation in the case of a zero value of initial deformation of structure often arises from such models. Obtained numerical data shows that initial orientation corresponding to a zero value of prolongation (δ = 0) is o_{init} in the considered tetrakaidecahedronal model. Since zero orientation is expected at zero deformation, the next modification of the degree of structure orientation was suggested:

$$o_s^{meas}(\delta) = \frac{o_s - o_{init}}{1 - o_{init}}. \tag{21}$$

We point out that we have not yet investigated the relationship between o_s^{meas} and the parameter o_{avr} in detail. Formula (21) is still our "preliminary hypothesis." But certainly the difference between o_s^{meas} and o_{avr} results from the model's limitations. However if the standard mechanism of plastic deformation is considered we believe that the linear relationship between o_s^{meas} and parameter o_{avr} is the acceptable option. The results obtained numerically using the formula (20) is shown in the Figure 9.

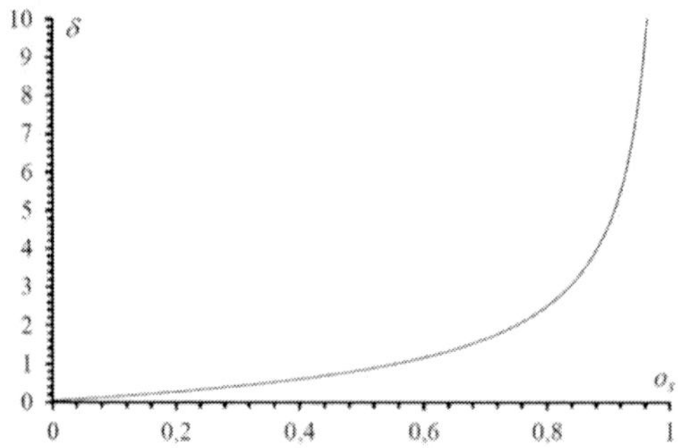

Figure 9. Elongation δ of the whole tetrakaidecahedronal grained structure as a function of the degree of structure orientation o_{avr}.

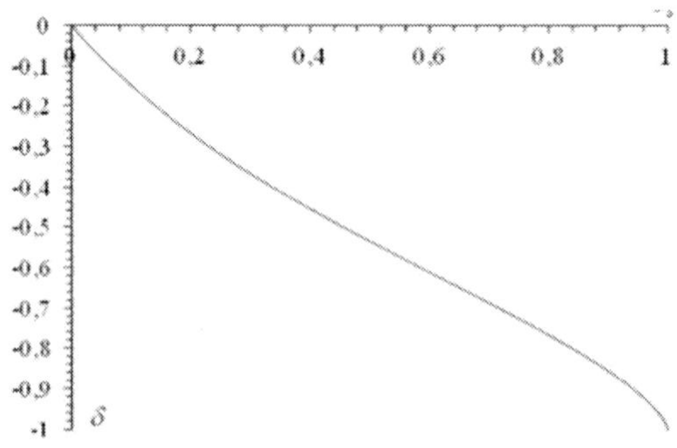

Figure 10. Contraction δ of the whole tetrakaidecahedronal grained structure as a function of the degree of structure orientation o_{avr}.

This corresponds with the concept of planar orientation. However the principle of calculation is exactly the same and was used in determining the graph in Figure 10. The average value of the degree of single grain orientation in the case of a contraction was calculated as:

$$o_s(\delta) = \frac{1}{4\pi^3} \int_0^{\pi} \int_0^{2\pi} \int_0^{2\pi} o(\varphi, \vartheta, \theta, \delta)\, d\theta\, d\varphi\, d\vartheta \tag{22}$$

and obtained values were corrected using the formula (21).

The numerical data calculated for the linear orientation of the structure using formula (20) and for the planar orientation of structure using formula (22) are summarized in Table 1A and Table 2A in the Appendix.

3. EXPERIMENTAL

The cold drawn tubes from S355J0 (1.0570) steel were analysed (the same as in [23]). The scheme of deformation is shown in Figure 10a, $\varphi_3 \neq \varphi_2$, planar orientation is observed in the plane perpendicular to φ_1 and from it the difference φ_3 and φ_2 can be determined, linear orientation is observed in the plane perpendicular to φ_3 or φ_2 (whichever is greater) and from this the value of φ_1 is determined. All three true strains were calculated form obtained values. The semi-product for cold drawing seamless tubes was a hot rolled tube (recrystallization was passed, grain boundaries deformation was minimized) with the following dimensions: outside diameter 70mm, wall thickness 6,3 mm, length 4000 mm.

The tube dimensions after deformation were as follows: outside diameter 50mm, wall thickness 3,75mm, length 9255mm. True macroscopic deformation was calculated from these dimensions (see Table 1). The probe was cut from the tube. The section plains were oriented in three main directions – parallel (perpendicular to φ_3), orthogonal (perpendicular to φ_1) and tangential (perpendicular to φ_2) in relation to the main tube's axes. The cuts were metallographically prepared - mechanically grinded and polished, chemically etched in 3% HNO_3

alcohol solution. On these plains the structure of the steel material was observed with the proper magnification of a light microscope. An example of the steel structure in the initial state and in the state after deformation is shown in Figure11.

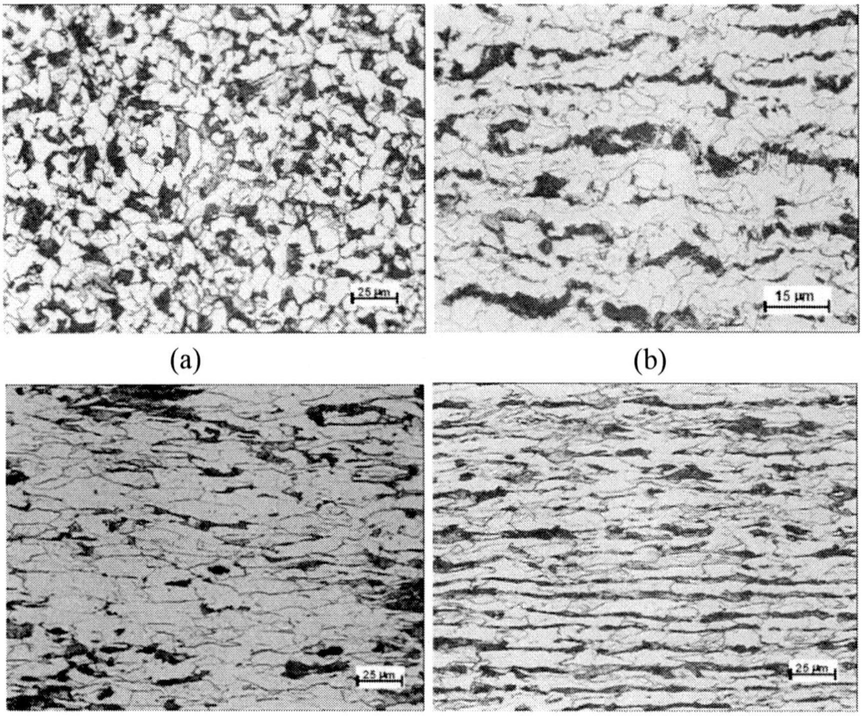

Figure 11. An example of structure in the middle of wall thickness: undeformed in the initial state a) and after deformation, b) in the orthogonal plane, c) in the tangential plane, d) in the parallel plane.

Table 1. The measured of the grain boundary's orientation and the calculated deformation according to the results in Figure 9 and Figure 10

Place	Deformed state						True macroscopic deformation		
	O_P	φ_2-φ_3	O_L	φ_1	φ_2	φ_3	φ_1	φ_2	φ_3
Out	0,103	-0,166	0,645	0,867	-0,546	-0,351	-	-	-0,337
Mid	0,212	-0,329	0,694	0,975	-0,652	-0,323	0,839	-0,519	-0,320
In	0,238	-0,364	0,7	0,993	-0,678	-0,315	-	-	-0,301

Degree of orientation of grain boundaries was measured in three places across the wall thickness – near the outside surface (Out), in the middle (Mid) and near the inside surface (In). Test lines were placed perpendicular and parallel to the grain boundary's orientation direction which was affected by straining (see Figure 3). From the specific number (number to unit of length) of parallel test line intersections with grain boundaries $(P_L)_P$ and perpendicular ones $(P_L)_O$ the linear orientation O_L of grain boundaries was calculated from equations (4) and (5), planar orientation O_P of grain boundaries was calculated from equations (2) and (3). The relative measurement precision was always smaller than 10% with reliability 90%. True strain φ_1 and the difference between true strains φ_2 and φ_3 was determined using the procedure described by equations (20), (22) and (21). True strain φ_2 and true strain φ_3 were calculated from values of true strain φ_1, true strains φ_2 - φ_3 difference and invariability of volume- $\varphi_1 + \varphi_2 + \varphi_3 = 0$. The results of the measurement of grain boundaries orientation in different places of the steel specimens and the calculated true deformations according to the results in Figure 9 and 10 are shown in Table 1.

CONCLUSION

This paper presents the application of the stereological technique on the modeling of local plastic deformation in grained structures. A major motivation of this paper was to describe observed changes of the grain's orientation, experimentally evaluated by the statistical Saltykov degree of orientation method, using deformation parameters. We mean strictly geometric orientation of the grain in three-dimensional space in our work. We do not discuss orientation of the crystal structure within the grain even though the concept of grain orientation is usually related to crystallographic orientation. The correlation between the degree of grain orientation and its deformation parameters was searched while the random orientation of undeformed grains was assumed. From the presented results we can conclude that our approach provides promising application possibilities.

The conversion model applied in this work takes into account all directions of possible deformation of idealized grains in the structure. This is closer to the real deformation than the model used in our previous work [23] where only one direction of deformation was preferred. The morphological anisotropy of polycrystalline material caused by plastic deformation can be described in this way. However the idealized shape of the grains was used in both cases.

A reasonable question to ask next is: what kind of model of the grained structure should be used in analyzing local deformations in specific materials? The model used will probably depend on the character of the particular grained structure. Systematic analysis of a great number of experimental data is necessary to find the answer to this question. Our results provide a good basis for the design of a comprehensive empirical model describing the kinetics of the grained structure deformation. Finally we would like to point out that this contribution is only intended to point out some preliminary aspects of the applicability of the Saltykov method in the description of local deformation process and such it could be useful in other theoretical models and simulations.

ACKNOWLEDGMENTS

This work was supported by the Slovak Research and Development Agency under the contract No. APVV-15-0319, by the VEGA Grant No. 1/0122/16 of the Grant Agency of the Slovak Republic Ministry of Education.

This work was funded by the ERDF - Research and Development Operational Programme under the project "University Scientific Park Campus MTF STU - CAMBO" ITMS: 26220220179.

REFERENCES

[1] Maier, B, Mihailova, B, Paulmann, C, Ihringer, J, Gospodinov, M, Stosch, R, Güttler, B, Bismayer, U. Effect of local elastic strain on

the structure of Pb-based relaxors: A comparative study of pure and Ba- and Bi-doped PbSc0.5Nb0.5O3. *Physical Review B* 2009 79, 224108.

[2] Stoudt, MR, Levine, LE, Creuziger, A, Hubbard, JB. The fundamental relationships between grain orientation, deformation-induced surface roughness and strain localization in an aluminum alloy. *Materials Science and Engineering A, 201* 530, 107 - 116.

[3] Wang, XG, Witz, JF, El Bartali, A, Dufrénoy, P, Charkaluk, E. Investigation of grain-scale surface deformation of a pure aluminium polycrystal through kinematic-thermal full-field coupling measurement. In: *13th International Conference on Fracture June 16–21*, Beijing China, 2013, 1 -10.

[4] M. P. Luong. Fatigue limit evaluation of metals using an infrared thermographic technique. *Mechanics of Materials*, 28(1998) 155–163.

[5] Moscicki, M, Kenesei, P, Wright, J, Pinto, H, Lippmann, T, Borbély, A, Pyzalla, AR. Friedels-pair based indexing method for characterization of single grains with hard X-rays. *Materials Science and Engineering A*, 2009 524, 64 - 68.

[6] Deng, J, Rokkam, S. A Phase Field Model of Surface-Energy-Driven Abnormal Grain Growth in Thin Films. *Materials Transactions*, 2011 52, 2126 – 2130.

[7] Gerth, D, Schwarzer, RA. Graphical Representation of grain and Hillock orientations in annealed Al-1%Si films. *Textures and Microstructures,* 1993 21, 77- 93.

[8] Sack, RA. Indirect Evaluation of Orientation in Polycrystalline Materials. *Journal of polymer science*, 1961 51, 543-560.

[9] Nian Zhang, Wei Tong. An experimental study on grain deformation and interactions in an Al-0.5%Mg multicrystal. *International Journal of Plasticity,* 2004 20, 523–542.

[10] Lewis, HD, Walters, KL, Johnson, KA. Particle size distribution by area analysis: Modifications and extensions of the Saltykov method. *Metallography,* 1973 6, 93–101.

[11] Xu, YH, Pitot, HC. An improved stereologic method for three-dimensional estimation of particle size distribution from observations in two dimensions and its application. *Comput Methods Programs Biomed,* 2003 72, 1-20.

[12] Rayaprolu, BD, Jaffrey, D. Comparison of discrete particle sectioning correction methods based on section diameter and area. *Metallography*, 1982 15, 193–202.

[13] Zhao, Z, Ramesh, M, Raabe, D, Cuitiño, AM, Radovitzky, R. Investigation of three-dimensional aspects of grain-scale plastic surface deformation of an aluminum oligocrystal. *International Journal of Plasticity,* 2008 24, 2278–2297.

[14] Patdar, V. A Note on geometry of grain boundaries. *Scripta Metalurgica*, 1986 20, 1227-1230.

[15] Bate, PS., Hutchinson, WB. Grain boundary area and deformation. *Scripta Materialia*, 2005 52, 199-203.

[16] Song, KJ, Wei, YH, Dong, ZB, Ma, R, Zhan, XH, Zheng, WJ, Fang, K. Constitutive model coupled with mechanical effect of volume change and transformation induced plasticity during solid phase transformation for TA15 alloy welding. *Applied Mathematical Modelling,* 2015 39, 2064–2080.

[17] Sharen, J, Cummins, P, Cleary, W. Using distributed contacts in DEM. *Applied Mathematical Modelling*, 2011 35, 1904–1914.

[18] Singh, SB, Bhadeshia, HKDH. Topology of grain deformation. *Materials Science and Technology*, 1998 14, 832-834.

[19] Saltykov, SA. *Stereometričeskaja metallografia*. (*Stereometric metallography*). Moskva: Metallurgia, 1970.

[20] Riosa, PR, Glicksmanb, ME. Modeling Polycrystals with Regular Polyhedra. *Materials Research*, 2006 9, 231-236.

[21] Chae, JY, Qin, R, Bhadeshia, HKDH. Topology of the Deformation of a Non-uniform Grain Structure. *ISIJ International*, 2009 49, 115–118.

[22] Zhu, Q, Sellars, CM, Bhadeshia, HKDH. Quantitative metallography of deformed grains. *Materials Science and Technology*, 2007 23, 757-775.

[23] Martinkovic, M, Minarik, S. Short notes on the grains modification by plastic deformation. In: Hubbard D. (Ed.), *Plastic Deformation.* New York: NOVA publishers, 2016, 1-44.

[24] Martinkovic, M, Pokorny, P. Estimation of local plastic deformation in cutting zone during turning. *Key Engineering Materials* 2015 662, 173 - 176.

APPENDIX

Table 1A. Obtained numerical data for grained structure elongation. Find the value of linear orientation determined by the Saltykov method in the last column of the table and assign to this value a local deformation parameter δ from the first column

Deformation (elongation) of structure *(δ)*.	Linear orientation of structure o_s calculated according the formula (20)	Linear orientation of structure modified according the formula (21). Corresponds to the graph in the Figure 9.
0	0,030618064	0
0,1	0,114892807	0,086936572
0,2	0,190551262	0,16498471
0,3	0,258112603	0,234679986
0,4	0,318290283	0,296758386
0,5	0,371858673	0,352018741
0,6	0,4195776	0,401244877
0,7	0,462153733	0,445165784
0,8	0,500224103	0,484438611
0,9	0,534352179	0,519644627
1	0,56503063	0,551292061
1,1	0,592687259	0,579822229
1,2	0,617692142	0,605616895
1,3	0,640364871	0,629005745
1,4	0,660981373	0,650273422
1,5	0,679780056	0,669665864
1,6	0,696967207	0,687395874
1,7	0,712721665	0,703647939

Table 1A. (Continued)

Deformation (elongation) of structure *(δ)*.	Linear orientation of structure o_s calculated according the formula (20)	Linear orientation of structure modified according the formula (21). Corresponds to the graph in the Figure 9.
1,8	0,727198822	0,718582359
1,9	0,740534031	0,732338762
2	0,752845499	0,745039089
2,1	0,764236734	0,756790118
2,2	0,774798621	0,767685604
2,3	0,784611178	0,777808092
2,4	0,793745045	0,787230453
2,5	0,802262747	0,796017188
2,6	0,810219775	0,80422554
2,61	0,81098659	0,805016576
2,7	0,8176655	0,81190644
2,8	0,824643957	0,819105312
2,9	0,831194517	0,825862772
3	0,837352465	0,83221522
3,1	0,843149495	0,838195349
3,2	0,848614136	0,843832593
3,3	0,853772129	0,849153502
3,4	0,858646742	0,85418208
3,5	0,863259051	0,858940069
3,6	0,867628186	0,863447203
3,7	0,871771538	0,867721424
3,8	0,875704953	0,871779077
3,9	0,879442888	0,875635074
4	0,882998558	0,879303051
4,1	0,886384063	0,882795487
4,2	0,889610498	0,88612383
4,3	0,892688055	0,889298592
4,4	0,895626106	0,892329442
4,5	0,898433286	0,895225287
4,6	0,901117558	0,897994342
4,7	0,90368628	0,900644197
4,8	0,906146253	0,90318187
4,9	0,90850378	0,905613859

Deformation (elongation) of structure (δ).	Linear orientation of structure os calculated according the formula (20)	Linear orientation of structure modified according the formula (21). Corresponds to the graph in the Figure 9.
5	0,910764702	0,907946193
5,1	0,912934442	0,910184464
5,2	0,915018039	0,912333872
5,3	0,917020184	0,914399255
5,4	0,918945245	0,916385119
5,5	0,920797294	0,918295665
5,6	0,922580133	0,920134816
5,7	0,924297315	0,921906235
5,8	0,925952161	0,92361335
5,9	0,927547783	0,925259369
6	0,929087094	0,9268473
6,1	0,930572827	0,92837996
6,2	0,932007549	0,929859998
6,3	0,93339367	0,9312899
6,4	0,934733456	0,932672003
6,5	0,93602904	0,934008509
6,6	0,937282432	0,935301489
6,7	0,938495524	0,936552896
6,8	0,939670101	0,937764573
6,9	0,940807851	0,938938259
7	0,941910365	0,940075596
7,1	0,942979148	0,941178136
7,2	0,944015623	0,942247349
7,3	0,94502114	0,943284625
7,4	0,945996972	0,944291279
7,5	0,946944331	0,94526856
7,6	0,947864364	0,946217652
7,7	0,948758158	0,947139677
7,8	0,949626748	0,948035702
7,9	0,950471116	0,948906739
8	0,951292194	0,949753751
8,1	0,952090871	0,950577654
8,2	0,952867991	0,95137932
8,3	0,95362436	0,952159579
8,4	0,954360742	0,95291922
8,5	0,955077869	0,953658997

Table 1A. (Continued)

Deformation (elongation) of structure *(δ)*.	Linear orientation of structure o_s calculated according the formula (20)	Linear orientation of structure modified according the formula (21). Corresponds to the graph in the Figure 9.
8,6	0,955776436	0,954379629
8,7	0,956457109	0,955081801
8,8	0,957120521	0,955766167
8,9	0,957767279	0,956433352
9	0,95839796	0,957083954
9,1	0,959013118	0,957718542
9,2	0,959613282	0,958337662
9,3	0,960198958	0,958941837
9,4	0,960770631	0,959531565
9,5	0,961328763	0,960107326
9,6	0,961873799	0,960669577
9,7	0,962406165	0,961218758
9,8	0,962926268	0,961755289
9,9	0,963434501	0,962279574
10	0,963931238	0,962792001

Table 2A. Obtained numerical data for grained structure contraction. Find the value of planar orientation determined by the Saltykov method in the last column of the table and assign to this value a local deformation parameter δ from the first column

Deformation (contraction) of structure *(δ)*	Planar orientation of structure o_s calculated according the formula (22)	Planar orientation of structure modified according the formula (21). Corresponds to the graph in the Figure 10.
0	0,024476269	0
-0,01	0,030245763	0,005914253
-0,02	0,036122907	0,011938857
-0,03	0,04210946	0,018075615
-0,04	0,04820718	0,024326329
-0,05	0,054417823	0,030692799

Deformation (contraction) of structure (δ)	Planar orientation of structure os calculated according the formula (22)	Planar orientation of structure modified according the formula (21). Corresponds to the graph in the Figure 10.
-0,06	0,060743136	0,037176817
-0,07	0,067184859	0,043780165
-0,08	0,073744719	0,050504615
-0,09	0,080424426	0,057351918
-0,1	0,087225674	0,064323812
-0,11	0,09415013	0,071422005
-0,12	0,101199438	0,078648183
-0,13	0,108375208	0,086003996
-0,14	0,115679018	0,093491062
-0,15	0,123112404	0,101110954
-0,16	0,130676856	0,108865201
-0,17	0,138373816	0,116755281
-0,18	0,14620467	0,124782614
-0,19	0,154170743	0,132948559
-0,2	0,162273292	0,141254404
-0,21	0,170513501	0,149701363
-0,22	0,178892474	0,158290567
-0,23	0,187411228	0,167023061
-0,24	0,196070688	0,175899789
-0,25	0,204871675	0,184921597
-0,26	0,213814903	0,194089214
-0,27	0,22290097	0,203403253
-0,28	0,232130346	0,212864199
-0,29	0,241503372	0,222472397
-0,3	0,251020243	0,232228051
-0,31	0,260681007	0,242131206
-0,32	0,270485549	0,252181748
-0,33	0,280433586	0,262379386
-0,34	0,290524658	0,272723646
-0,35	0,300758114	0,283213863
-0,36	0,311133106	0,293849168
-0,37	0,321648579	0,304628478
-0,38	0,332303259	0,315550489
-0,39	0,343095645	0,32661366
-0,4	0,354023997	0,337816209
-0,41	0,365086329	0,349156099

Table 2A. (Continued)

Deformation (contraction) of structure *(δ)*	Planar orientation of structure o_s calculated according the formula (22)	Planar orientation of structure modified according the formula (21). Corresponds to the graph in the Figure 10.
-0,42	0,376280397	0,36063103
-0,43	0,387603687	0,372238426
-0,44	0,399053411	0,383975428
-0,45	0,410626494	0,395838884
-0,46	0,422319564	0,407825338
-0,47	0,434128946	0,419931022
-0,48	0,446050653	0,432151849
-0,49	0,458080375	0,444483401
-0,5	0,470213475	0,456920925
-0,51	0,482444981	0,469459324
-0,52	0,494769579	0,482093152
-0,53	0,50718161	0,494816605
-0,54	0,51967506	0,507623521
-0,55	0,532243561	0,520507371
-0,56	0,544880385	0,533461257
-0,57	0,557578443	0,546477914
-0,58	0,570330282	0,559549702
-0,59	0,583128086	0,572668608
-0,6	0,595963676	0,585826248
-0,61	0,608828513	0,599013869
-0,62	0,621713697	0,612222347
-0,63	0,634609974	0,625442197
-0,64	0,64750774	0,638663572
-0,65	0,660397043	0,651876273
-0,66	0,673267594	0,665069751
-0,67	0,686108771	0,678233118
-0,68	0,698909627	0,691355153
-0,69	0,711658899	0,704424309
-0,7	0,724345011	0,71742872
-0,71	0,736956087	0,730356214
-0,72	0,749479959	0,743194314
-0,73	0,761904166	0,755930249
-0,74	0,774215967	0,768550958

Deformation (contraction) of structure *(δ)*	Planar orientation of structure o_s calculated according the formula (22)	Planar orientation of structure modified according the formula (21). Corresponds to the graph in the Figure 10.
-0,75	0,786402339	0,781043091
-0,76	0,798449983	0,793393015
-0,77	0,810345314	0,805586805
-0,78	0,822074464	0,817610243
-0,79	0,833623261	0,829448804
-0,8	0,844977219	0,841087638
-0,81	0,856121508	0,852511541
-0,82	0,867040915	0,86370492
-0,83	0,877719801	0,874651743
-0,84	0,888142026	0,885335466
-0,85	0,898290871	0,895738949
-0,86	0,908148913	0,905844334
-0,87	0,917697882	0,91563289
-0,88	0,926918457	0,925084813
-0,89	0,935790009	0,934178956
-0,9	0,944290247	0,942892468
-0,91	0,952394748	0,951200313
-0,92	0,960076292	0,959074591
-0,93	0,967303916	0,966483559
-0,94	0,974041484	0,973390174
-0,95	0,980245425	0,979749775
-0,96	0,985860899	0,985506143
-0,97	0,990814635	0,99058417
-0,98	0,994999533	0,99487407
-0,99	0,998231963	0,998187603
-1	1	1

In: Polycrystalline Materials
Editor: Jim Olson

ISBN: 978-1-53613-864-1

Chapter 3

Electric Current Assisted Sintering of Ceramics – Steps and Pitfalls on the Way to Transparency

Vojtěch Nečina and Willi Pabst*

Department of Glass and Ceramics, University of Chemistry and Technology, Prague (UCT Prague), Prague, Czech Republic

Abstract

Electric current assisted sintering (ECAS), also called pulsed electric current sintering (PECS) or field assisted sintering technique (FAST), but more widely known under the popular – albeit less appropriate – name spark plasma sintering (SPS), has gained considerable attention from the scientific community during the last few decades, because it is a convenient method for the fast densification of otherwise hard-to-sinter ceramic powders, both oxide and non-oxide. While in the case of metal powders the densification is always affected by the flow of electric current through the sample itself (initially via the particle contact points and subsequently via the sinter necks), this is usually not the case for ceramic powders, at least not for temperatures below which ionic

* Corresponding Author Email: pabstw@vscht.cz.

conductivity plays a significant role. Therefore, in contrast to metals, in ceramics it is essentially only the Joule heating of the mold (usually a graphite mold) and the ensuing high heating rate that distinguishes ECAS from other forms of (uniaxial) hot pressing. Thus, also the sintering mechanisms in ECAS are not too different from those of (uniaxial) hot pressing. Since ECAS is an extremely versatile method, allowing the simultaneous control of many process parameters and conditions, such as electric current levels and pulse rates, temperature schedules (including heating and cooling rates, dwell times and hold temperatures), pressure schedules (including pressure increase and decrease rates, dwell times and hold pressures) and atmospheres (usually vacuum or inert), the number of actual and potential applications of ECAS is steadily increasing and already now too large to be treated in a single review. Therefore, the present contribution focusses on recent ECAS achievements and progress in the special field of translucent and transparent ceramics. For the preparation of these materials appropriate grain size, extremely high density and the minimization (preferably absence) of second phase inclusions are critical factors. Following an outline describing the historical development of ECAS and related techniques, the principle of ECAS is briefly described, current opinions concerning sintering mechanisms are summarized, and the problem of carbon contamination is discussed in some detail. In this context the use of boron nitride as a protective coating and lithium fluoride as a sintering aid is mentioned. Fundamental optical properties are defined, an overview of microstructural factors affecting transmittance is given, and literature data (ECAS conditions and resulting optical properties) are compared for various oxide (and a few silicate, phosphate or non-oxide) ceramics that have so far been prepared in transparent or at least translucent form for wavelengths close to the visible or infrared range of the electromagnetic spectrum. The materials reviewed in this chapter include, among others, alumina, lutecia, magnesia, yttria, zirconia, titanates, zirconates, lutetium compounds, magnesium aluminate spinel (magnesia alumina spinel), yttrium aluminum garnet (YAG), hydroxyapatite, mullite, aluminum nitride, aluminum oxynitride, silicon aluminum oxynitride, calcium fluoride, magnesium fluoride, sphalerite, and yttria-magnesia nanocomposites.

Keywords: electric current assisted sintering (ECAS), spark plasma sintering (SPS), field assisted sintering technique (FAST), pulsed electric current sintering (PECS), sintering temperature, heating rate, pressure, hot pressing, dwell time, grain size, pore size, porosity, transparent ceramics, transparency, translucent materials, in-line transmittance, Mie scattering theory, refractive index, absorption, wavelength (ultraviolet / UV, visible, infrared / IR), optical properties, discoloration, aluminum oxide (alumina), lutetium oxide (lutecia), magnesium oxide (magnesia), yttrium oxide (yttria), zirconium oxide (zirconia), titanate, zirconate, lutetium compounds, perovskite, pyrochlore, magnesium aluminate spinel (magnesia alumina spinel), yttrium aluminum garnet (YAG), phosphate, hydroxyapatite, silicate, apatite structure, mullite, aluminum nitride, aluminum oxynitride, silicon aluminum oxynitride, calcium fluoride, magnesium fluoride, sphalerite, yttria-magnesia nanocomposite

1. Introduction

Electric current assisted sintering (ECAS), also called pulsed electric current sintering (PECS) or field assisted sintering technique (FAST), but more widely known under the popular – albeit less appropriate – name spark plasma sintering (SPS), has gained considerable attention from the scientific community during the last few decades, because it is a convenient method for the fast densification of otherwise hard-to-sinter ceramic powders, both oxide and non-oxide. While in the case of metal powders the densification is always affected by the flow of electric current through the sample itself (initially via the particle contact points and subsequently via the sinter necks), this is usually not the case for ceramic powders, at least not for temperatures below which ionic conductivity plays a significant role. Therefore, in contrast to metals, in ceramics it is essentially only the Joule heating of the mold (usually a graphite mold) and the ensuing high heating rate that distinguishes ECAS from other forms of (uniaxial) hot pressing. Thus, also the sintering mechanisms in ECAS are not too different from those of (uniaxial) hot pressing. Since ECAS is an extremely versatile method, allowing the simultaneous control of many process parameters and conditions, such as electric current levels and pulse rates, temperature schedules (including heating and cooling rates, dwell times and hold temperatures), pressure schedules (including pressure increase and decrease rates, dwell times and hold pressures) and atmospheres (usually vacuum or inert), the number of actual and potential applications of ECAS is steadily increasing and already now too large to be treated in a single review. Therefore, the present contribution focusses on recent ECAS achievements and progress in the special field of translucent and transparent ceramics. It may read as complementary reading and an update to other authors' excellent previous reviews on ECAS in general (Garay 2010, Guillon et al. 2014, Munir et al. 2006, Orrù et al. 2009) and the application of ECAS to transparent ceramics (Kodera et al. 2013, Wang et al. 2013).

2. Electric Current Assisted Sintering (ECAS) – Principle and Historical Development

Electric current assisted sintering (ECAS) is a versatile sintering technique that can be used for electrically conducting materials (e.g., metals and alloys) as well as electrically non-conducting materials such as ceramics, which are usually insulating at room temperature and may become conductive only at high temperatures (e.g., due to ionic conductivity). The principle of ECAS consists in the direct heating of the electrically conducting mold, and in the case of metal powders also the powder compact itself, via Joule heating. This mode of heating, usually performed via direct current (DC) pulses, can provide extremely high heating rates, commonly up to several hundred °C/min, in extreme cases up to 1000°C/min (Jiang et al. 2008, Quach et al. 2010), which are rates much higher than those attainable in uniaxial hot pressing (HP) or isostatic hot pressing (HIP), which are of order 10°C/min. Thus, compared to HP and HIP, ECAS provides a more efficient way to fabricate dense polycrystalline materials, offering energy savings that result from the direct mode of heating (of the mold, in contrast to the convective-radiative heating in conventional sintering and hot pressing) and higher throughput rates due to the higher heating rates and shorter processing times, with which HP or HIP cannot compete.

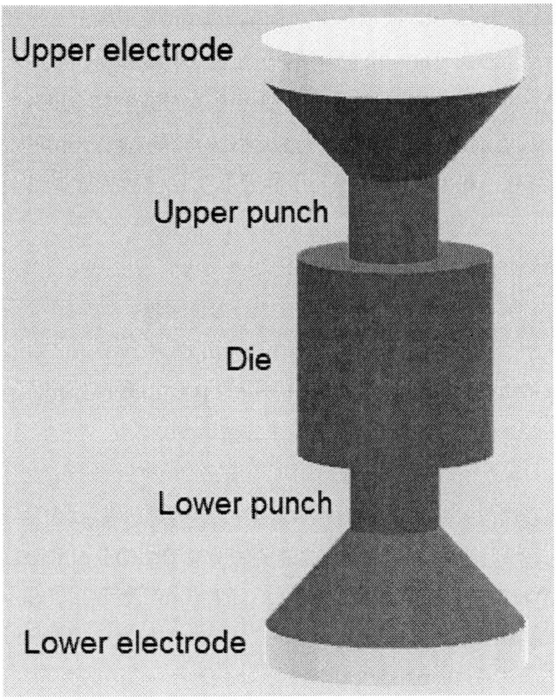

Figure 1. Schematic view of an assembled ECAS mold (die and punches).

The mold (die) is at the same time used as a pressing device, so that the intrinsic driving force of sintering (i.e., the sintering stress due to the curvature of the particles) is enhanced by external pressure, thus allowing efficient densification at lower temperature without the interference of too much grain growth. This feature of grain growth suppression is principally similar to HP and HIP, but may be further enhanced due to the higher heating rates. Although several mold materials have been used, including tungsten carbide (WC), silicon carbide (SiC) and hardened steel (up to 600°C (Quach et al. 2010)), graphite is the most common one, see Figure 1. In vacuum or inert atmospheres it can be used at temperatures up to at least 2400°C. The maximum pressures that common graphite molds can withstand are around 100 MPa, but special mold constructions using WC or SiC punches have been reported to withstand pressures even in the range 500–1000 MPa (Ghanizadeh et al. 2017, Sokol et al. 2017). Common pressures used in ECAS are below 100 MPa (Quach et al. 2010).

The voltage across an electrically non-conductive powder compact is of order 100 V/m (Quach et al. 2010), which is several orders of magnitude below the breakdown voltages of ceramic materials (order of magnitude MV/m (Kingery et al. 1976)), but the electric current flowing through the mold are relatively high, *viz*. around 1–5 kA in small laboratory and up to 60 kA in large industrial ECAS equipment (Quach et al. 2010). Although it has been claimed by some authors that very intense DC pulses might produce sparks, i.e., localized plasma discharges, in the small gaps between the grains of the powder (Tokita 1993), and clean grain boundaries are sometimes assumed to be due to these plasma discharges, evidence has always been indirect and inconclusive. Actually, more recent systematic research (Hulbert et al. 2008) has concluded that plasma formation in ECAS is highly unlikely, so that the popular acronym SPS is surely a misnomer. Also the fact that current pulses are used instead of a continuous current seems not to be very important, at least not for electrically insulating materials, such as ceramics. Therefore, also the acronym PECS is not very useful. A similar critique refers to the witty term FAST, because even in electrically non-conductive powder compacts the electric field across the sample is negligibly small, if not completely by-passed by the electric current flowing through the die mantle, while in electrically conducting powder compacts a field is not established at all, because the current flows though the compact itself.

The historical origins of ECAS have to be sought in the early patent literature, in which the use of electric current in sintering processes was proposed (Bloxam 1906a, 1906b, Cremer 1944, Duval d'Adrian 1922, Taylor 1932, 1933a, 1933b), see also (Grasso et al. 2009, Munir et al. 2006, Orrù et al. 2009). The development of ECAS continued, under different names, during the 1950s, 1960s and 1970s, with milestones in this development being Inoue's 1966 and 1967 patents (Inoue 1966a, 1966b, 1967a, 1967b), but it is only in the last few decades, after Inoue's patents had expired in the late 1980s (Grasso et al. 2009), that commercial ECAS equipment has become more readily available for research and development, as well as for industrial production. The worldwide development of instrumental equipment was led by Japanese companies (mainly Sumitomo Coal Mining Co., Ltd), and also research in the field of ECAS was and still is heavily dominated by Japan and its neighbor countries China and Korea (Munir et al. 2006, Orrù et al. 2009). For more details on the historical development and current trends of ECAS the interested reader may refer to very detailed reviews (Grasso et al. 2009, Orrù et al. 2009).

Table 1. Scientometric data for titles and topics related to electric current assisted sintering (ECAS) since 1945 (according to the Thomson-Reuters Web of Science Core Collection, accessed 13 March 2018); the usage of the acronyms FAST, PECS is difficult to estimate (because it collides with the use of the words "fast" and "PECs" in other contexts), and topics with SPS are so common that a scientometric evaluation is not available (n/a)

Title or topic	Searched item	Total number of publications	Average citations per paper	H-index (Hirsch index)	Total number of citations
Title	Spark plasma sintering	3909	11.72	73	45817
Title	SPS	2248	10.68	68	24015
Title	Field assisted sintering	141	14.27	24	2012
Title	Pulsed electric current sintering	178	13.38	25	2382
Title	Electric current assisted sintering	15	52.00	5	780
Title	ECAS	4	42.00	2	168
Topic	Spark plasma sintering	8059	11.30	94	91087
Topic	SPS	19365	n/a	n/a	n/a
Topic	Field assisted sintering	599	14.12	41	8456
Topic	Pulsed electric current sintering	489	13.10	35	6408
Topic	Electric current assisted sintering	141	19.29	22	2720

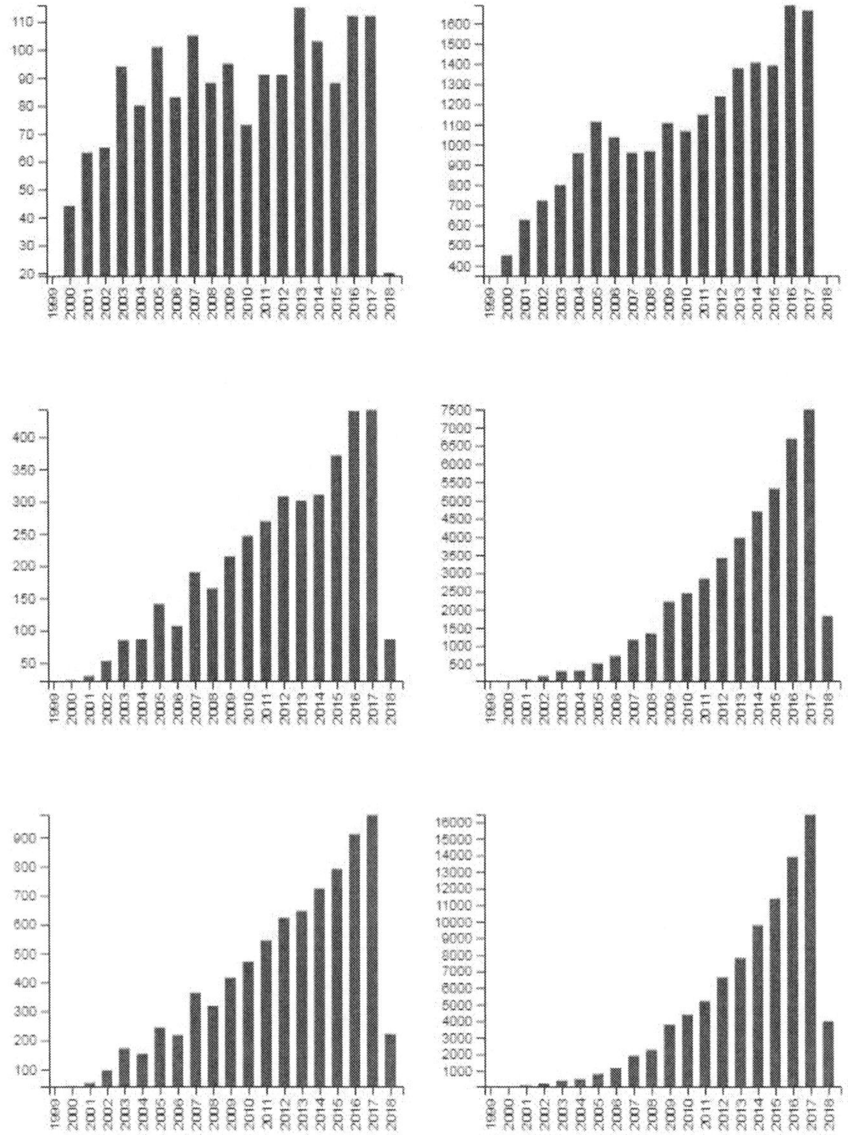

Figure 2. Publications (left) and citations (right) with "SPS" in the title (top row), "spark plasma sintering" in the title (middle row) and "spark plasma sintering" in the topic (bottom row); source: Thomson-Reuters Web of Science Core Collection (accessed 13 March 2018).

The scientometric data in Table 1 indicate that most authors prefer the term "spark plasma sintering" or its acronym "SPS." That means, albeit misleading, the term cannot be eliminated from scientific-technical tradition any more, whereas the term preferred by the present authors, "electric current assisted sintering" or "ECAS," albeit the most adequate one, is the least used one. However, interestingly, the average number of citations per paper is much higher for papers using the latter, which might be one of the indicators of quality and scope. The term "field assisted sintering technique" is as unsuitable as the term "pulsed electric current sintering"/"PECS," because actually neither the field nor the pulses are essential to the process, but obviously the former, apart from being heavily propagated by the leading German producer of this type of equipment (FCT Systeme GmbH), has the obviously attractive acronym "FAST," which definitely captures a key feature of the process (the other important key feature being the application of pressure). A look on the dynamics of scientific activity in the field, see Figure 2, shows an approximately linear increase of publications and an exponential increase of citations since the year 2000, which is obviously closely related to the wider availability of appropriate equipment from commercial producers.

3. Sintering Mechanisms and Special Issues Related to ECAS of Transparent Ceramics

The driving force of ECAS is very similar to conventional hot pressing techniques (HP, HIP), i.e., the sintering stress due to the surface curvature of the convex particles is significantly enhanced by the external uniaxial pressure. Also the principal sintering mechanisms can be expected to be essentially the same as in more conventional sintering techniques. In particular, in ceramics the sintering mechanisms typical for solid state sintering, i.e., surface, grain boundary and volume (lattice) diffusion, will be supported by high-temperature creep, i.e., either grain-boundary-diffusion-based Coble creep or volume-diffusion-based Nabarro-Herring creep (Rahaman 2003, 2010). The enhancement of the diffusion rates by the electric field, i.e., electromigration (Munir et al. 2006, Quach et al. 2010), is highly improbable, because the electric field strengths are comparably low. Other possible but disputable effects of electric field, electric current and pulsing are reviewed in (Munir et al. 2006, Quach et al. 2010).

Since electric current assisted sintering (ECAS) of electrically non-conductive materials, typically ceramics, is essentially an efficient and fast hot pressing technique, the sintering driving forces, stages and mechanisms are very similar to more conventional sintering techniques. Then main difference is the possibility to achieve very high heating rates, which allow densification to start at lower temperature (Johnson 1990, Zhou et al. 2004). This would be an advantage, for example, if grain growth could be efficiently suppressed. On the other hand, in modern sintering theory, mainly in the context of master sintering curves, it is usually assumed that grain size cannot be decoupled from densification (Su & Johnson 1996). Actually, from the viewpoint of sintering theory it can be expected that by fast heating, together with the applied pressure, the initial stage of sintering, in which surface diffusion may be dominant, is by-passed to some degree, thus using what has been called a "kinetic window" (Eriksson et al. 2011).

In contrast to ECAS of electrically conductive powders, typically metals, where the electric current through the compact sample itself can lead to local heating at the contact points and to enhanced neck formation kinetics (Frei et al. 2007, Song et al. 2006, Yanagisawa et al. 2003), these effects can be largely ignored in ceramics, because at temperatures at which some ceramics may become ionically conductive the necks will be large enough for local effects to be negligible. However, the influence of thermodiffusion due to the temperature gradients within the sample (Ludwig-Soret effect), may be an issue (Olevsky & Froyen 2009, Quach et al. 2010) and has been invoked by some authors (Quach et al. 2010) to explain the fact that – although higher heating rates are generally favorable for densification – a better final density is achieved when the heating rate in the final stage of sintering is not too high (Shen et al. 2002), an effect that can be of special importance for transparent ceramics (Morita et al. 2009).

Transparent materials fabricated by ECAS often display discoloration, which is attributed mostly to oxygen vacancies, polyvalent ions, and / or carbon contamination (from the graphite die, more precisely the graphite paper between the powder compact and the die mantle wall and punch faces). In the majority of cases various shades of gray, black, yellow, brown and orange coloration are reported. Most authors explain this discoloration as being a consequence of the vacuum or reducing environment present during processing in the ECAS equipment, which leads in structural defects, mostly oxygen vacancies, in the powder compact, see Figure 3. Oxygen vacancies absorb electromagnetic radiation at wavelengths lower than 400 nm (Morita et al. 2012), which is a wavelength region in which most ionic solids (except for earth alkaline earth fluorides, for example, see below) are not very transparent anyway.

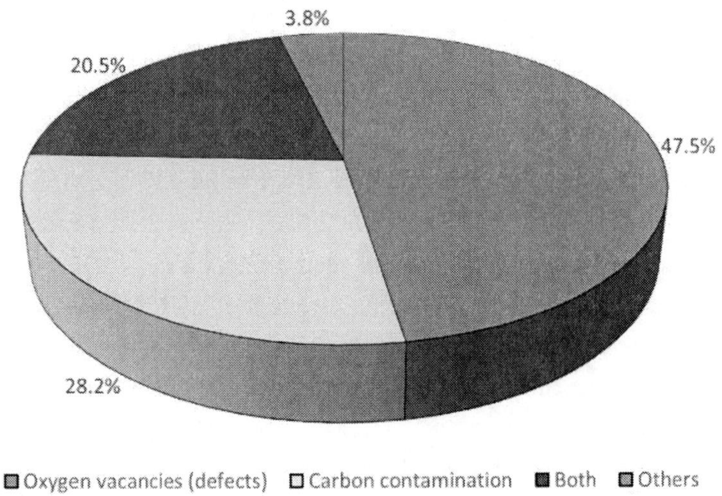

Figure 3. Percentage of tentative explanations of the discoloration in transparent ceramics fabricated by ECAS, according to the literature.

These defects are usually eliminated by so-called post-annealing, performed in air at elevated temperatures for a sufficiently long time. In most cases the transmittance measured is higher after post-annealing than before (over a wide range of wavelengths), see below. Moreover, when polyvalent ions, i.e., ions that can exist in multiple valence states, are present, the valence state of the dopant may change. Due to the reducing environment of ECAS, these ions tend to adopt a state of lower valency, resulting often in absorption at certain wavelengths, i.e., a narrow absorption band in the visible part of the electromagnetic spectrum. Also such discoloration can be removed by post-annealing in air via re-oxidation of the ion into a higher valency state. Actually this problem and its solution is not specific to ECAS, but occurs commonly also in other contexts of vacuum sintering. For example, in transparent yttrium aluminum garnet (YAG) ceramics doped with rare earth elements for solid state laser applications, the valency state of some of these dopants may change, e.g., from Yb^{3+} to Yb^{2+}, during vacuum sintering, which is accompanied by the occurrence of oxygen vacancies. Post-annealing in this case re-establishes the required (laser-active) valency (Yb^{3+}) by re-oxidation and at the same time eliminates the oxygen vacancies. At the same time the green color of the material disappears, so that the final product is a colorless transparent ceramic (Esposito et al. 2014, Hostaša et al. 2012, 2014a, 2014b). It should be mentioned that some authors assume that the decrease of transparency which is sometimes observed after post-annealing and usually attributed to an increase in porosity is caused by the coalescence of oxygen vacancies (Zhang et al. 2010, 2011c, Fu et al. 2014). However, considering this question as definitely answered would seem premature.

Carbon contamination is a specific problem of ECAS, because graphite molds are most commonly used, usually in connection with graphite paper. Since the graphite die or paper is in direct contact with the powder compact, it has been assumed in early work that carbon is directly diffusing into the sample during the ECAS process. However, diffusion of carbon through the lattice of the sintered material is highly unlikely, if not impossible.[1] Therefore, Morita and coworkers (Morita et al. 2015a, 2015b, 2015c) proposed that the carbon evaporates from the graphite die or paper, due to the high temperature and vacuum, so that gaseous carbon enters the

[1] Actually, even the self-diffusion coefficient of carbon in graphite is extremely low (Kingery et al. 1976).

powder compact via open pores where it condenses, so that it is subsequently entrapped in the closed pores and forms a more or less structured carbon phase. It has been proposed that in the case of low contamination, the carbon in amorphous, whereas in the case of high contamination, the carbon phase tends to be more structured (Morita et al. 2015c). Indeed, it seems plausible that carbon that has evaporated from plane or low-curvature surfaces (i.e., the punch faces or the die wall) may condense (at the same temperature) in the vicinity of high-curvature surfaces, such as contact points or sinter necks between particles. This mechanism is similar to the classical evaporation-condensation mechanism in solid state sintering (Kingery & Berg 1955, Kingery et al. 1976). Solid carbon itself could account for the wide range absorption observed before post-annealing, see below. Moreover, after post-annealing some authors (Ahmadi et al. 2016, An et al. 2011a, 2011d, Irom et al. 2016, 2018, Kim et al. 2014) observed an absorption peak at a wavelength of approximately 4250 nm, which is clearly related to gaseous carbon monoxide (CO) and carbon dioxide (CO_2) and indicates oxidation of carbon in the closed pores. Actually, the CO/CO_2 absorption peak emerges (after post-annealing) in many cases, although some authors tend to ignore it or are not aware of it (Ahasanzadeh-Vadeqani & Shoja Razavi 2016, Du et al. 2017, Ghaderi et al. 2016, Jiang et al. 2008, Shoja Razavi et al. 2016, Zhang et al. 2011b). In earlier work some authors (An et al. 2011a, 2011d) have assumed that CO and CO_2 are trapped in the pore space during densification. However, there is little justification for such a mechanism, since vacuum is present during the whole ECAS process and the CO/CO_2 absorption peak is observed only after post-annealing in air. In the same way as oxygen vacancies are eliminated by post-annealing, because at the high temperature the oxygen diffusion coefficients in oxide ceramics are high, also the carbon in closed pores is oxidized. According to the Boudouard equilibrium $CO_2 + C \rightarrow 2CO$, which is shifted towards the right hand side when the temperature is high (at 1 atm gas pressure the equilibrium temperature is 705°C), CO is produced during post-annealing, which partly decomposes into CO_2 and solid carbon during cooling (Salmang & Scholze 1982). The problem with this carbon oxidation in closed pores is pore swelling and the concomitant increase of porosity.[2] This is obviously the reason why some authors observed a decrease in transmittance after post-annealing, even though the discoloration was removed, especially when the annealing step was performed at high temperature. Post-annealing can cause pore swelling not only due to gas formation as such or the thermal expansion of the gas, but also due to the shift of the Boudouard equilibrium to the right hand side, i.e., toward the formation of carbon monoxide, so that at high temperature the gas pressure on the pore wall is relatively high. In principle, also the coalescence of pores by Ostwald ripening (Rahaman 2003) might explain an increase in porosity, because smaller closed pores (with higher surface curvature) contain gas at a higher pressure (and thus smaller volume) than larger pores. Therefore, the coalescence of pores would increase not only pore size, but also porosity. However, even at high temperature this mechanism is highly unlikely, because of the insolubility of both CO and CO_2 in the solid. While in early works reduced transmittance has been attributed either exclusively to oxygen vacancies or exclusively to carbon contamination, in more recent works usually both causes are mentioned.[3] It has been pointed out, however, that the discoloration is weaker when lower heating rates are used in the ECAS process (Morita et al. 2015b, 2015c). This supports the theory that carbon contamination is responsible for the discoloration rather than oxygen vacancies, because oxygen vacancies will be created preferentially during high temperature and prolonged time, whereas in the case of high heating rate the discoloration is stronger, although the creation of oxygen vacancies is much more difficult. Moreover, the discoloration is usually weaker, when the pre-load applied during ECAS is lower (Morita et al. 2012, Wang & Zhao 2009). Interestingly, carbon contamination is not

---

[2] Naturally, both CO and CO_2 try to occupy much more space than the solid carbon. Moreover, both gases are virtually insoluble in the solid, so that they cannot escape from the closed pores, because their diffusion coefficients in the solid are practically zero.

[3] Of course, also impurities in the starting material can be the source of discoloration. For instance, organic or carbonate residues can be transformed into carbon at high temperatures in reducing environments, which is typical for ECAS. Inorganic impurities, but also dopants that have been intentionally added, can segregate at the grain boundaries, especially when the ion diameter is too large to fit into the host lattice, and coalesce into larger inclusions during ECAS. This is generally undesirable, because these inclusions can act as scattering centers.

observed for samples prepared via ECAS at low temperatures (below approximately 1000°C).

Another effect is the non-uniform heating the powder compact is not heated uniformly, when high heating rates are applied, because heating occurs essentially by conduction in the radial direction from the die wall to the center of the sample.[4] This will result in preferential sintering and earlier pore closure at the sample periphery (i.e., usually the edge of the circular disk) compared to the sample center, preventing gases evolving at high temperatures, e.g., SiO (Salmang & Scholze 1982), and the carbon that has entered the compacts as long as the porosity was open, from escaping. This can finally result in a non-uniform distribution of porosity, with a dense rim and larger, gas- or carbon-filled, pores in the center of the specimen.

Although not always explicitly mentioned by authors, graphite paper or foil is commonly used in ECAS in order to avoid direct contact and possible reactions of the powder compact with the die wall and punch faces, see Figure 4. A thinkable possibility to avoid or at least reduce the carbon contamination may consist in coating the die mantle wall and punch faces with a layer of (hexagonal) boron nitride (Li et al. 2016a). Although this type of coating is not often explicitly mentioned in the open literature concerning transparent ceramics, this procedure is relatively well known and some authors have doubtlessly tried it. However, the results are rather disappointing so far. Albeit promising at first sight, attempts by the present authors to use boron nitride coatings for reducing the carbon contamination in the preparation of magnesium aluminate spinel via ECAS at temperatures in the range 1400–1500°C were actually not successful.[5]

Another potential approach to reduce the carbon contamination is replacing the graphite paper by another type of foil, e.g., a metal foil. Nearly carbon-free transparent magnesium aluminate spinel ceramics have

---

[4] The temperature of the punches can be expected to be lower than that of the die mantle, because the electrical current, by flowing through the die mantle, essentially by-passes the punches. Heating in the axial direction is expected to be slower, because for electrically non-conductive powder compacts the electric current does not flow through the powder compact itself, but compared to the radial temperature gradient the axial temperature gradient is usually negligible because of the small thickness of the powder compact.

[5] In vacuum at such high temperatures boron nitride tends to decompose.

been fabricated using platinum foil wrapped around the powder compact (Biswas et al. 2016). In another study molybdenum foil was used to prepare transparent calcium fluoride (Wang et al. 2017b), also with promising results. Interestingly, however, it was observed that molybdenum foil not only prevents carbon contamination, but at the same supports grain growth. Average grain sizes of approximately 16 μm and 260 μm were reported for transparent calcium fluoride when using graphite paper and molybdenum foil, respectively. The authors (Wang et al. 2017b) explain this significant difference by the clean grain boundaries in the latter case. Indeed, similar to other secondary phase inclusions, also carbon inclusions are expected to act as barriers to grain boundary migration, thus inhibiting grain growth. While small grain size is always an advantage from the viewpoint of mechanical properties, it can be favorable also for optical properties, especially in the case of anisotropic crystallites (Kim et al. 2012). In particular, a grain size smaller than the wavelength of light or electromagnetic radiation for which transparency is required will usually guarantee that the size of the intergranular voids (pore size) is small enough not to impair transparency.

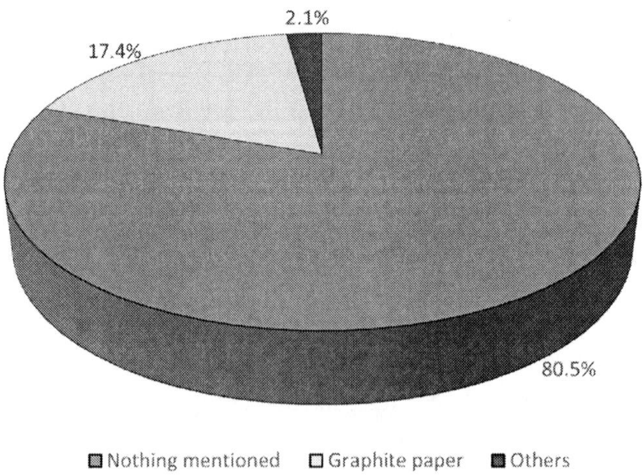

Figure 4. Percentage of materials used by different authors as intermediate layers between the powder compact and the die wall and punch faces used in ECAS according to the literature.

In the majority of studies concerning the fabrication of transparent ceramics by ECAS, as summarized in this chapter, small amounts of sintering aids are used. While in traditional ceramic technology, sintering aids often create liquid melt phases that remain in the ceramic after sintering and subsequent cooling in the form of a glassy (amorphous) or crystalline phase (or phase mixture) at the grain boundaries (Rahaman 2003), in the case of transparent ceramics it must be guaranteed that the liquid melt phase, if present at higher temperatures, is only a transient one and does not result in the formation of secondary phases that would impair transparency.

With the exception of alumina, where magnesia (MgO) is frequently utilized on traditional grounds, the most commonly used sintering aid in ECAS is lithium fluoride (LiF). Its effect is still not quite clear, although several studies were performed in order to clarify its role. In the case of magnesia-aluminate spinel, which is one of the most thoroughly investigated transparent ceramics it has been concluded (Rozenburg et al. 2007, 2008, Reimanis & Kleebe 2007, Müller & Kleebe 2012) that molten LiF starts to act as a transient liquid around 800°C. At higher temperature (around 1000°C) the surface of the spinel particles starts to dissolve in the LiF melt, leaving behind in a defective structure with oxygen vacancies. Upon further heating, LiF starts to leave the system and the dissolved spinel reprecipitates. Moreover, it is assumed that LiF acts as cleansing agent and forms volatile species with gaseous carbon according to the equation (Meir et al. 2009).

$$n\,\mathrm{LiF} + n\,\mathrm{C} \rightarrow n\,\mathrm{Li} + (\mathrm{CF})_n. \qquad (1)$$

The disadvantage of LiF is that, due to this cleansing effect, it promotes grain growth and thus deteriorates some of the mechanical properties.

4. Optical Properties and Microstructural Factors Affecting Transmittance

A material, more precisely a slab of material (i.e., a disk- or plate-like specimen) with given thickness, can be called transparent when a sharp image is obtained by a view through, no matter how large the distance is between the observer and the object viewed, see Figure 5. This definition of transparency, which is similar to that given by other authors (Krell et al. 2009a, 2009b) and could be extended beyond the visible part of the electromagnetic spectrum, distinguishes transparent materials from translucent materials such as Lucalox® alumina, which provide an increasingly blurred image with increasing distance. Of course, another question is the clarity (brightness) of the image, because each material has a maximum transparency that is a consequence of its intrinsic electromagnetic material properties, in particular the dielectric function, and cannot be exceeded.

Transmittance is a quantitative measure of the transparency of a solid material. Experimentally, the degree of transparency is determined by transmission measurements, usually via photospectrometry. Measured values are reported as in-line transmittance, real in-line transmittance or total forward transmittance, but some authors abstain from precisely defining their results and report them just as "transmittance" or "transmission." Needless to say, that this practice can lead to misinterpretations and misunderstandings. While the total forward transmittance (commonly measured using a so-called integrating sphere) contains all intensity contributions that leave the specimen on the opposite side and thus can be registered at angles smaller than 90°, no matter whether they have been scattered or not, the in-line transmittance is the transmittance only in the direction of the incident light[6] beam or ray (in terms of geometrical optics). Of course, this condition can be fulfilled only approximately, because in order to fulfil it exactly the aperture opening

[6] Here and in the sequel it is understood that the word "light" stands in a broader sense for all types of electromagnetic radiation.

angle and thus the aperture diameter in front of the detector would have to approach zero, so that the intensity reaching the detector would approach zero as well. Therefore, strictly speaking, in-line transmittance results should always be accompanied by information on the aperture opening angle. While in common photospectrometers these angles are usually a few degrees (3–5°), modern precision equipment can readily achieve aperture opening angles of 0.5°. The latter value has therefore been conventionally used for defining the so-called real in-line transmittance (Apetz & van Bruggen 2003).

Figure 5. Transparent spinel ceramics prepared via ECAS; note that transparency is defined by sharp images of objects in large distance, not just by the visibility of objects directly in contact with the rear side of the specimen.

The absolute transmittance[7] determines the intensity (more precisely, the irradiance) of light that is able to propagate through a specimen (usually a slab, i.e., a disk or plate) of a given thickness. This absolute transmittance cannot be higher than the theoretical maximum transmittance, which is uniquely determined by the optical properties of

[7] Here and in the sequel we are talking about the ideal in-line transmittance, which is approximately realized (and probed) via real in-line transmittance measurements.

the material, i.e., the (possibly complex) index of refraction, and the surface state of the specimen (i.e., the surface roughness, depending on which reflections of incident light can be specular or diffuse). The reader may find more detailed explanations in several excellent monographs (Bohren & Huffman 2004, van de Hulst 1981) or in our recent papers (Pabst & Hostaša 2013, Pabst et al. 2014). When absorption is absent (i.e., when wavelengths of light far away from the absorption bands of the solid materials are considered and thus the absorption coefficient is negligibly small, approximated to be zero) and the surface is polished in a way to exhibit perfectly specular reflection, the reflectance R corresponding to the degree of reflection, i.e., the percentage of reflected light intensity, at a single surface for perpendicular incidence of light from vacuum, air or any other gas into the solid material, is given via the Fresnel equation

$$R = \left(\frac{n-1}{n+1} \right)^2 , \tag{2}$$

where n is the (real part of the) refractive index of the solid material. The light that is not reflected enters the solid material and is reflected again at the opposite surface. At this surface, part of the light leaves the slab in the original direction, but part of it is reflected again and thus propagates in the opposite direction, where upon incidence on the surface part of it is reflected again etc. As a result of these multiple reflections, the total reflectance, i.e., the total percentage of reflection losses of the light transmitted in the direction incidence, is

$$R_{total} = \frac{2R}{1+R} . \tag{3}$$

In the absence of absorption and scattering, the theoretical maximum transmittance in the exact forward direction, i.e., the maximum real in-line transmittance in the original direction of the incident light beam, is then the fraction of light that has not been reflected, i.e.,

$$T_{\max} = 1 - R_{total} = \frac{2n}{1+n^2}. \quad (4)$$

Of course, since the refractive index is a function of the wavelength, also the maximum transmittance is wavelength-dependent. Table A1 in Appendix A lists refractive indices of materials that have been prepared via ECAS and are treated in this chapter. Note that, while cubic crystals and crystallites, as well as the polycrystalline materials prepared from them, are optically isotropic, non-cubic crystals are optically anisotropic, so that the calculation of the effective refractive index of polycrystalline materials consisting of randomly oriented anisotropic crystallites has to be based on some kind of averaging procedure. Based on the fact that for non-magnetic (diamagnetic) materials the refractive index at a given wavelength is uniquely determined by the relative permittivity (dielectric constant) at that wavelength, i.e.,

$$n = \sqrt{\varepsilon_r}, \quad (5)$$

and the fact that the effective dielectric constant of a dense(pore-free) polycrystalline material with randomly oriented crystallites is approximately given by the arithmetic mean of the three principal permittivities of the individual crystallites, i.e.,

$$\varepsilon_r = \frac{\varepsilon_{r1} + \varepsilon_{r2} + \varepsilon_{r3}}{3}, \quad (6)$$

the effective refractive index of these dense polycrystalline materials in terms of the principal refractive indices of the anisotropic crystallites is approximately given by the relation

$$n = \sqrt{\frac{n_1^2 + n_2^2 + n_3^2}{3}}. \quad (7)$$

This relation has been used to calculate the maximum transmittances listed in Table A1 for the polycrystalline materials consisting of anisotropic crystallites. Of course, the lower the birefringence (i.e., the difference between the principal refractive indices), the closer are the optical properties of the individual crystallites to those of cubic and thus optically isotropic crystals, rendering this averaging procedure redundant.

Of course, the experimentally measured absolute transmittance can never exceed the theoretical maximum transmittance. Actually it is always lower because of either absorption or scattering or both. The combined effect of scattering and absorption is called extinction. Via extinction the intensity (irradiance) of the light beam (ray) in the exact forward direction is attenuated while propagating through the specimen (slab of material). This attenuation of light intensity (from I_0 for the incident light to I for the transmitted light) increases with the thickness of the specimen. It is known that the dependence of this attenuation on the specimen thickness is an exponential one, as described by the Lambert-Beer law. Defining the relative transmittance as the ratio of the absolute and maximum transmittance, which is at the same time the ratio between the transmitted and incident intensity of light (irradiance), the Lambert-Beer law is

$$T_{rel} = \frac{T}{T_{\max}} = \frac{I}{I_0} = \exp\left(-\gamma_{ext} h\right), \tag{8}$$

where I is the transmitted intensity of light, I_0 the incident intensity of light, h the thickness of the specimen (slab) and γ_{ext} the extinction coefficient, which in general contains both scattering and absorption. For the case of absorption the reader is referred to (Bohren & Huffman 2004, van de Hulst 1981). In the absence of absorption the extinction coefficient reduces to the scattering coefficient, see (Pabst & Hostaša 2013, Pabst et al. 2014). Note that both the absolute and the relative transmittance depend on the specimen thickness (therefore transmittance information without information on the specimen thickness is generally useless), while the extinction or scattering coefficient does not. The latter may therefore be considered as a certain kind of effective material property. Scattering occurs in general at optical heterogeneities and particularly at high-phase-contrast inclusions, i.e., inclusions with a refractive index that is significantly different from that of the surrounding medium (matrix). This is the case for vacuum, but also air or other gases, i.e., media with refractive indices of 1.0, whereas solids usually have refractive indices in the range 1.4–2.7 (Pabst & Hostaša 2013). The higher the phase contrast the stronger the scattering. Therefore the presence of pores (vacuum voids or gas-filled) always leads to scattering, i.e., in terms of geometrical optics to the deflection of light away from the forward direction, so that the in-line transmittance decreases. Slightly different is the situation for strongly absorbing inclusions, i.e., carbon or graphite inclusions, which play an important role in materials fabricated by ECAS. Even though the real part of the complex refractive index is closer to that of the matrix in this case, the finite absorption coefficient (or absorption index, i.e., the imaginary part of the complex refractive index) lead to a strong decrease of transmittance (not only the in-line transmittance, but the total forward transmittance), see below.

For spherical inclusions the extinction or scattering coefficient can be calculated using Mie theory or its approximations for small (Rayleigh) or large (Fraunhofer) size compared to the wavelength of light (Bohren & Huffman 2004, van de Hulst 1981). Actually a combination of the Rayleigh and Fraunhofer approximation can in some cases be used to obtain useful approximations to Mie theory, as has been demonstrated in the context of our so-called R-scaling approach (Pabst & Hostaša 2013, Pabst et al. 2014).

In general, it may be said that the transmittance is a very sensitive function of the concentration (volume fraction) of inclusions, so that transparency can only be expected for fully densified materials. This is a realistic target for ECAS processing. Indeed, when appropriate ECAS conditions are chosen, the porosity in ECAS-processed ceramics can be extremely low, not only because of the combination of temperature and (uniaxial) pressure, but also because of the vacuum that usually prevails during ECAS. On the other hand, carbon contamination is a major problem in ECAS, and it is known that these carbon or graphite inclusions can completely destroy transparency. Therefore, the influence of these inclusions has to be taken into account in any realistic assessment of using the ECAS process for the preparation of transparent ceramics.

Table 2. Input data for the Mie scattering calculations to obtain the transmittance of yttrium aluminum garnet (YAG) with pores or carbon (graphite) inclusions

Medium	Refractive index at 600 nm	Refractive index at 1064 nm	Absorption index at 600 nm	Absorption index at 1064 nm
YAG	1.832	1.815	-	-
Carbon (graphite)	1.673	1.890	0.889	1.276
Vacuum	1.000	1.000	-	-

Figures 6 through 9 show the pore or inclusion size dependence of the transmittance of yttrium aluminum garnet (YAG) ceramics with porosities (pore volume fractions) or inclusion volume fractions of 10^{-6} and 10^{-5}, calculated via Mie scattering theory for vacuum wavelengths of 600 nm and 1064 nm, corresponding to visible light and infrared (IR) radiation from a Nd-YAG laser, respectively. All calculations are for ideally polished, plane-parallel specimens with a thickness of 1 mm. Also shown are the theoretical maximum transmittances of YAG for the corresponding wavelengths. The input data for these calculations, which have been obtained via Polyanski's Refractive Index Database (Polyanskiy 2018) for undoped YAG (Zelmon et al. 1998) and Asbury micro graphite (Querry 1985), are given in Table 2. The Mie scattering calculations themselves have been performed using Prahl's Mie Scattering Calculator (Prahl 2018). Calculations for carbon inclusions without absorption are purely hypothetical and have been performed only for the purpose of comparison with vacuum voids or gas-filled pores, i.e., to show the influence of absorption.

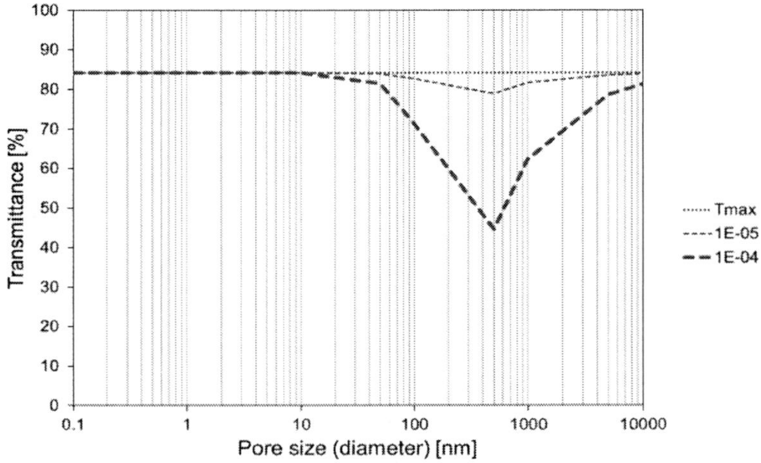

Figure 6. Pore size dependence of the transmittance of yttrium aluminum garnet (YAG) ceramics with porosities (pore volume fractions) of 10^{-5} and 10^{-4}, calculated via Mie scattering theory for a wavelength of 600 nm, corresponding to visible light (specimen thickness 1 mm); also shown is the theoretical maximum transmittance of YAG.

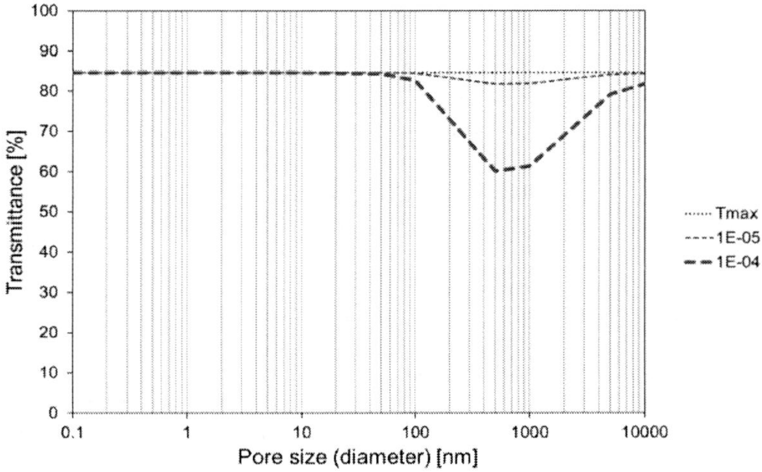

Figure 7. Pore size dependence of the transmittance of yttrium aluminum garnet (YAG) ceramics with porosities (pore volume fractions) of 10^{-5} and 10^{-4}, calculated via Mie scattering theory for a wavelength of 1064 nm, corresponding to IR radiation from a Nd-YAG laser (specimen thickness 1 mm); also shown is the theoretical maximum transmittance of YAG.

From Figures 6 and 7 it is evident that the transmittance decreases significantly even for extremely low volume fractions of pores, when the pore size is close to the wavelength of light or electromagnetic radiation used. In the examples calculated for specimens with thickness 1 mm, volume fractions of 0.0001, i.e., 0.01%, lead to transmittance minima of 44.6% and 60.1% at wavelengths of 600 nm and 1064 nm, respectively (compared to theoretical maximum transmittances of 84.1% and 84.5% at these wavelengths), when the pore size is between 500 and 1000 nm, i.e., close to the wavelength. The corresponding extinction coefficients are 0.0634 and 0.0342, respectively. This enormous influence of low volume fractions of pores is of course caused by the high phase contrast between pores (vacuous or gas-filled) and solid, which is 1 to 1.832 and 1.815, respectively. For the same volume fraction, pores smaller than 20 nm and 100 nm (and also pores larger than 10 μm) affect the transmittance at wavelengths of 600 nm and 1064 nm, respectively, only to a negligible degree.

As shown by the calculations of (hypothetical) non-absorbing inclusions with refractive index 1.673 or 1.890 in Figures 8 and 9, the influence for comparable volume fractions would be much smaller, resulting in transmittance minima of 72.0% and 79.8% at wavelengths of 600 nm and 1064 nm, respectively, and the transmittance minima are shifted to larger inclusion size (around 1000 nm and 5000 nm, respectively). However, due to the presence of absorption, which is an essential feature of carbon (graphite) inclusions, this picture is changed completely. While for inclusions larger than 10 μm the decrease in transmittance is relatively low (81.5% and 81.9%, respectively), it decreases steeply below 1 μm and rapidly approaches asymptotic limit values of 11.0% and 17.2%, respectively, for inclusion sizes below 100 nm. It may be concluded that in contrast to nanometric pores and other non-absorbing inclusions, nanometric absorbing inclusions are absolutely detrimental for transparency.

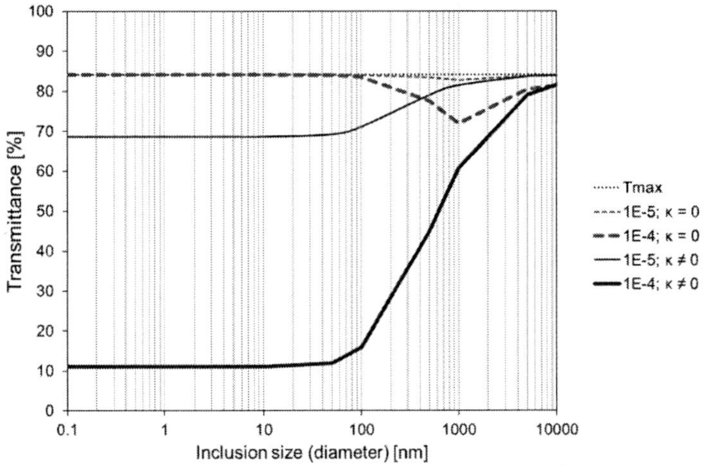

Figure 8. Inclusion size dependence of the transmittance of yttrium aluminum garnet (YAG) ceramics with carbon contamination volume fractions of 10^{-5} and 10^{-4}, calculated via Mie scattering theory for a wavelength of 600 nm, corresponding to visible light, with and without absorption coeffcients taken into account (specimen thickness 1 mm); also shown is the theoretical maximum transmittance of YAG.

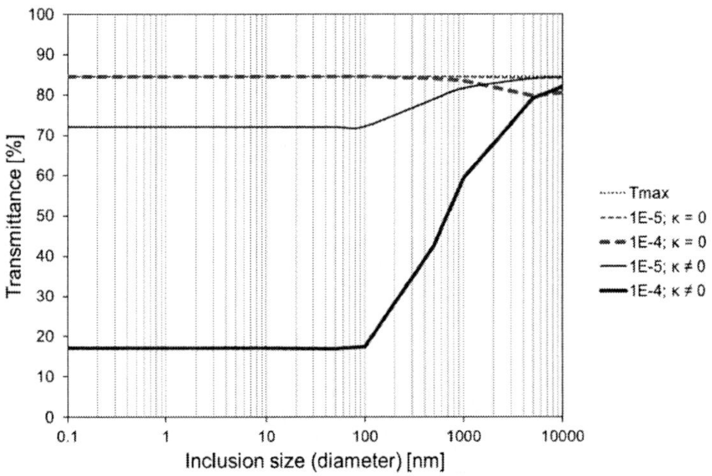

Figure 9. Inclusion size dependence of the transmittance of yttrium aluminum garnet (YAG) ceramics with carbon contamination volume fractions of 10^{-5} and 10^{-4}, calculated via Mie scattering theory for a wavelength of 1064 nm, corresponding to IR radiation from a Nd-YAG laser, with and without absorption coeffcients taken into account (specimen thickness 1 mm); also shown is the theoretical maximum transmittance of YAG.

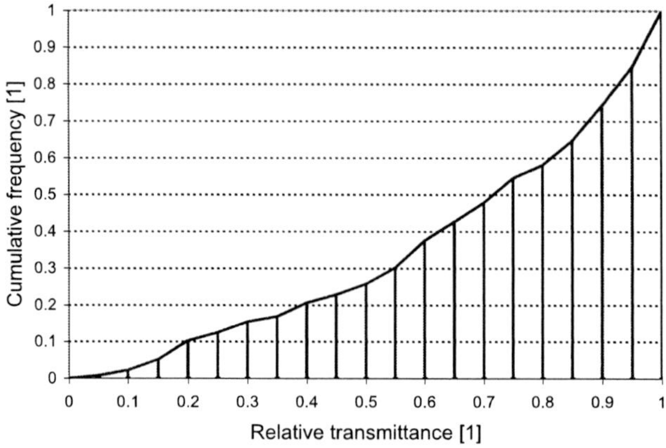

Figure 10. Cumulative frequency curve (= curve showing the cumulative sum of the number of optimal results, normalized to unity) of relative transmittances achieved for ECAS-processed ceramics designed for transparency (based on 136 literature values in the wavelength range 300–5680 nm); the median value corresponds to a relative transmittance of approximately 0.72 (= 72% of the theoretical maximum transmittance for a given material), i.e., in 50% of all cases a higher transmittance has been achieved, in 50% a lower one; note, however, that these values are based on spectrophotometric measurements of specimens with different thickness, ranging from 0.2 to 3.5 mm (average 1.2 ± 0.6 mm); that means, similar to the absolute transmittance, also the relative transmittance is not a material property, but a performance parameter related to a body of certain dimensions (here a specimen or slab of given thickness).

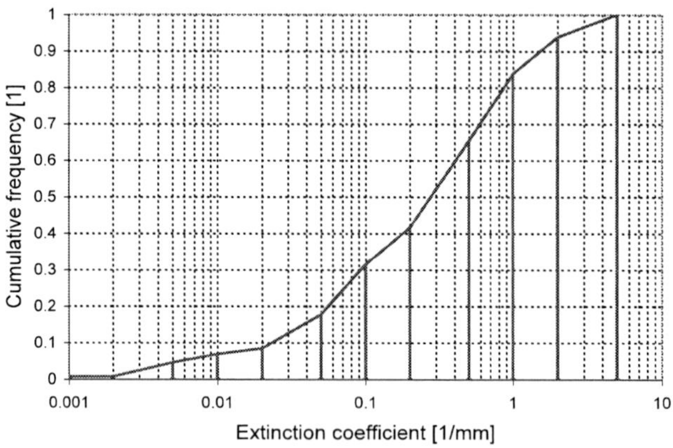

Figure 11. Cumulative frequency curve (= curve showing the cumulative sum of the number of optimal results, normalized to unity) of extinction coefficients achieved for ECAS-processed ceramics designed for transparency (based on 130 literature values in the wavelength range 300–5680 nm); the median value corresponds to an extinction coefficient of approximately 0.3, i.e., in 50% of the cases the extinction coefficient achieved is lower and in 50% of the cases it is higher than 0.3; note that, as long as the concentration of scattering centers in the material is low, extinction coefficients are independent of the sample thickness.

It has to be emphasized that all these model calculation results are for second-phase inclusions in homogeneous isotropic media (assumed to be non-absorbing). Ideally dense (i.e., pore-free) polycrystalline materials that are single-phase (i.e., without second phase inclusions) and consist entirely of cubic crystallites should exhibit the theoretical maximum transmittance, see Table A1. However, if the crystallites are non-cubic, the maximum transmittance cannot be achieved even in dense single-phase materials. The reason is that due to birefringence, the grain boundaries of differently oriented grains act as scattering centers as well. Therefore the actually attainable maximum transmittance in these materials is always lower than the theoretically calculated value listed in Table A1, even in the absence of second-phase inclusions. A widely used modeling approach for taking this birefringence into account has been proposed by Apetz and van Bruggen (Apetz & van Bruggen 2003). Although the assumptions of this approach are disputable (Pabst & Hostaša 2013), it has been extensively used for modelling the transmittance of transparent alumina and tetragonal zirconia ceramics (Klimke et al. 2011, Krell et al. 2009a, 2009b, Trunec et al. 2016).

It is well known that the value of the extinction coefficient depends (according to Mie's scattering theory) on the size of the inclusions, their concentration and the phase contrast between the inclusions and the matrix material.[8] That means, not even the extinction coefficient, albeit independent of the sample thickness, can be considered as a material property in the strict sense. However, for sufficiently small inclusions and inclusions with sufficiently high phase contrast, e.g., vacuum or gas-filled inclusions (closed pores) in solids or carbon/graphite inclusions in transparent ceramics, the Rayleigh and Fraunhofer approximations to Mie's theory apply. In these cases, the extinction coefficient can be considered as a relative material property, determining the transmittance of a piece of material, irrespective of the solid phase. In this case, which may hold at least as an approximation, the extinction coefficient would allow a direct, material-independent, comparison of the transparency results

[8] Theoretically also the shape of the inclusions plays a role, but a quantification of this factor is highly non-trivial and often not necessary.

achieved via ECAS, i.e., a direct assessment of the capability of different ECAS schedules to produce dense ceramic microstructures with a required degree of transparency.

It has to be recalled, however, that the conditions for this comparison to be admissible may change with the wavelength, so that conclusions obtained for one wavelength need not be automatically valid for a completely different wavelength. The cumulative frequency curves in Figures 10 and 11 represent collected data for different materials in the wavelength range 300–5680 nm, i.e., from the ultraviolet (UV) to the infrared (IR) with 1260 ± 1187 nm on the average (6 x below 500 nm, 26 x 500–550 nm, 62 x 600–650 nm, 31 x 680–1500 nm, 17 x 2000 nm and 24 x 2200–5680 nm). The selection of wavelengths used for the characterization of optical properties is of course dictated on the one hand by the availability of instrumental equipment, (e.g., photospectrometers), on the other hand by the intended application (e.g., IR transmittance for heat radiation detection etc.) However, with respect to the fact that the grain size and pore size is usually not well characterized, the attempt to analyze literature data with respect to differences in wavelength would be futile at the moment. It has to be recalled, however, that it is not the absolute size of the inclusions, but their relative size compared to the wavelength that determines the scattering regime (Rayleigh, Mie or Fraunhofer).

5. Comparison of Literature Data for ECAS Conditions and Optical Properties

This section gives a rather comprehensive and up-to-date overview of what has been achieved in the preparation of different potentially transparent ceramics prepared via ECAS. The section is subdivided into simple oxides, mixed oxides, phosphates and silicates, as well as non-oxides, oxynitrides and composites. Basic processing conditions, including the ECAS schedule (either the range tested or the optimum one or both), and optical properties (mainly the transmittances attained) are reportedwherever available. The transmittances have to be understood as absolute transmittances for the material in question and the specimen thickness and wavelength given. It is clear that these values cannot be directly compared, even if the specimen thickness and wavelength are identical, because different materials have different theoretical limit values for the maximum transmittance. Moreover, not all authors measure transmittance in a sufficiently small solid angle to deserve the name "real in-line transmittance," although most authors try to approach this case as closely as possible. Therefore the transmittance cited here is in most cases an "in-line transmittance." Notwithstanding these reservations in detail, the following overview should give a realistic overall picture of current achievements in the field and should provide a rough guide of viable ECAS processing conditions and schedules. In this section the results for individual materials are summarized. A more detailed listing of the results of individual papers by different authors is given in Appendix B.

5.1. Simple Oxides

Simple oxide ceramics that have been prepared by ECAS in transparent or translucent form are mainly alumina Al_2O_3 (trigonal), lutecia Lu_2O_3 (cubic), magnesia MgO (cubic), yttria Y_2O_3 (cubic) and zirconia ZrO_2 (both cubic and tetragonal), see Appendix B1. While translucent alumina ceramics has been developed in the late 1950s (Burke 1990) and has been used, under the trade name Lucalox®, since the 1960s for high-pressure sodium vapor lamps, which have become the most popular street lighting all over the world, really transparent alumina, usable for high-strength, scratch-resistant and impact-resistant optical windows and armor, is a much more recent development (Krell et al. 2009a, 2009b, Trunec et al. 2016). Actually, alumina is not predetermined for transparent applications, because it is not cubic, but the birefringence is relatively low so that the transmittance reduction due to scattering at grain boundaries is relatively limited (Apetz & van Bruggen 2003). The same is true for tetragonal zirconia, usually doped with 3 mol% Y_2O_3 (Cascolo et al. 2008, Klimke et al. 2011), which can be used for impact-resistant optical windows with increased fracture toughness. On the other hand, cubic zirconia, usually doped with 8 mol% Y_2O_3, and the other aforementioned cubic oxides are optically isotropic and therefore ideally suited for preparing transparent ceramics. While transparent magnesia can be used for infrared windows and dosimeters (Harris 2005, Kato 2016a, 2016b, 2016c, 2017b), yttria can be used for missile domes and, similar to lutecia, with adequate doping it is a suitable luminescence and scintillator material as well as a host material for ceramic solid state lasers (Harris 1998, Ikesue 2008, Liu et al. 2014, Petermann et al. 2002, Zhang et al. 2011c).

For alumina ceramics, which have been prepared in transparent form via ECAS by different authors (Alvarez-Clemares et al. 2010, 2012, Aman et al. 2009, 2010, Ghanizadeh et al. 2017, Grasso at al. 2010, 2011, Jiang et al. 2008, Jin et al. 2010, Kim et al. 2007, 2009a, 2009b, 2010b, 2012, 2016, Lallemant et al. 2012, Prakasam et al. 2016a, Roussel et al. 2011, Stuer et al. 2010, Suárez et al. 2013, Tõldsepp et al. 2016, Wang & Zhao 2010, Wang et al. 2014), the sintering temperatures range from 900 to 1680°C, but are typically in the range 1100–1500°C, with heating rates up to several hundred °C/min (even 1000°C/min in one extreme case), but typically around 100°C/min, maximum pressures ranging from 40 MPa to several hundred MPa (typically 80–100 MPa) and dwell times usually below 60 min (typically 20 min or less). For specimens with thickness of around 1 mm (0.8–1.0 mm) absolute transmittances up to 64% have been achieved in the visible region of the electromagnetic spectrum and 71–86% at infrared wavelengths. Post-annealing (at 1250°C for 24 h) has been used only in one case (Jiang et al. 2008).

For lutecia ceramics, which have been prepared in transparent form via ECAS by different authors (Alombert-Goget et al. 2015, An et al. 2011b, 2011c, 2011d, 2014a, 2014b, Boulesteix et al. 2014, Ito et al. 2016, Kijko et al. 2015, Kurosawa et al. 2014, Prakasam et al. 2013, 2014), the sintering temperatures range from 1000 to 1800°C, but are typically in the range 1400–1500°C, with heating rates up to 200°C/min, maximum pressures ranging from 20 to 130 MPa (typically 100 MPa) and dwell times usually in the range 5–45 min. For specimens with thicknesses in the range 1.0–2.0 mm absolute transmittances in the range 40–80% have been achieved in the visible region of the electromagnetic spectrum and 64–82% at infrared wavelengths. In many cases grey discoloration has been observed and post-annealing treatments (at temperatures in the range 850–1400°C for times in the range 2–12 h) have often been used to eliminate it.

For magnesia ceramics, which have been prepared in transparent form via ECAS by different authors (Chaim et al. 2004, Tran et al. 2012, Kato et al. 2016a, 2016b, 2016c, 2017b, Jiang et al. 2017, Kawano et al. 2017), the sintering temperatures range from 650 to 1500°C, but are typically in the range 1400–1500°C, with heating rates up to 300°C/min, maximum pressures ranging from 30 to 500 MPa (typically 80 MPa) and dwell times usually in the range 5–60 min. For specimens with thicknesses in the range 0.8–1.5 mm absolute transmittances in the range 16–37% have been achieved in the visible region of the electromagnetic spectrum and 71–77% at infrared wavelengths. Post-annealing has not been used.

For yttria ceramics, which have been prepared in transparent form via ECAS by different authors (Ahmadi et al. 2016, Ahsanzadeh-Vadeqani & Shoja Razavi 2016, An et al. 2012a, Attia et al. 2015, Du et al. 2017, Furuse et al. 2018, Ghaderi et al. 2016, Hu et al. 2017, Irom et al. 2016, 2018, Korkmaz & Sahin 2017, Shoja Razavi et al. 2016, Yoshida 2011, Zhang et al. 2011c), the sintering temperatures range from 850 to 1600°C, but are typically in the range 1050–1500°C, with heating rates up to 280°C/min (typically 100°C/min), maximum pressures from 45 to 300 MPa (typically 100 MPa) and dwell times usually in the range 5–60 min (typically 20–30 min). For specimens with thicknesses in the range 1.0–1.7 mm the absolute transmittances in the visible region are in a very wide range (6–75%), whereas transmittances at infrared wavelengths are up to 80–83% for specimens as thick as 2.5–3.5 mm. Post-annealing treatments (at temperatures in the range 800–1300°C for times in the range 1–10 h) have often been used.

In the field of advanced fine ceramics, two types of zirconia ceramics are commonly produced, viz. fully stabilized zirconia (FSZ), which consists entirely of cubic crystallites (c-ZrO_2), and tetragonal zirconia polycrystals (TZP), i.e., ceramics consisting entirely of tetragonal crystallites (t-ZrO_2). The former, usually doped with 8 mol% of Y_2O_3, is most commonly used for functional applications (e.g., as ionic conductors), while the second, usually doped with 3 mol% of Y_2O_3, is most commonly used as a structural material, because it is optimized for maximum strength and/or fracture toughness. Although for optical applications the former is intrinsically more suitable, because it consists of cubic crystallites, which are optically isotropic (i.e., free of birefringence), several attempts have also been directed towards the fabrication of transparent TZP ceramics. This is possible, because the crystallographic structure of t-ZrO_2 is relatively close to that of c-ZrO_2, so that also the birefringence, albeit present, is still relatively low. The fabrication of transparent TZP ceramics is a challenging task, because TZP ceramics have far better mechanical properties than FSZ ceramics. TZP ceramics also have certain advantages compared to alumina ceramics, in particular higher fracture toughness. Therefore, if the preparation of TZP in transparent form could be mastered, this would be a major breakthrough in the field of strong, impact-resistant optical windows. On the other hand, TZP ceramics are prone to hydrothermal aging, i.e., the low-temperature surface degradation in contact with aqueous media. This phenomenon, which is due to local yttria depletion, can be avoided by a small amount of alumina as an additional dopant.

For zirconia ceramics, which have been prepared in transparent form via ECAS by different authors (Alaniz et al. 2009, Lei et al. 2012, Prakasam et al. 2016b, Xiong et al. 2011, Zhang et al. 2010, 2011b, 2011d, 2012), the sintering temperatures range from 1000 to 1400°C, but are typically in the range 1100–1200°C, with heating rates up to 200°C/min, maximum pressures from 40 to 400 MPa and dwell times in the range 5–180 min (typically around 10 min). For cubic zirconia specimens (fully stabilized zirconia/FSZ) with thicknesses in the range 0.6–1.3 mm the absolute transmittances achieved were 28–56% at wavelengths of visible light, while for tetragonal zirconia polycrystals (TZP) a transmittance of 63% was obtained in the infrared region (3000 nm). However, alumina-doped TZP ceramic samples exhibited rather bad transparency, corresponding to a transmittance of only 7% in the visible region (640 nm) for a 1 mm thick specimen (Zhang et al. 2012). This may indicate a principal limitation in material optimization for transparent TZP ceramics. Although a small alumina content in yttria-doped zirconia ceramics has the advantage of increasing the resistance to hydrothermal ageing (low-temperature surface degradation due to yttria depletion in aqueous environments), it seems that it is not possible to achieve at the same time optimal optical properties (transparency). Post-annealing of zirconia ceramics prepared via ECAS has been performed at temperatures in the range 750–900°C for times 2–24 h. For transparent zirconia ceramics prepared via ECAS see also (Klimke et al. 2011).

Among the few other simple oxide ceramics that have been prepared via ECAS are calcium oxide (Kato et al. 2017a) and zinc oxide (Lin et al. 2017, Prakasam et al. 2014). The former (CaO), which may be applied as a phosphor or scintillator material (Yanagida 2013), has been prepared at 1600°C with a heating rate of 70°C/min, pressure 80 MPa and dwell time 60 min, but optical properties have not been given. The latter (ZnO), which is also a potential scintillator material (Gorokhova et al. 2008), has been prepared at temperatures in the range 850–1100°C with heating rates up to 100–200°C/min, pressures 40–100 MPa, dwell times 10–20 min (and in one case a post-annealing step at 700°C for 5 h). The best transmittance achieved was 69% at wavelengths of visible light for a specimen with thickness 2.0 mm (Prakasam et al. 2014).

5.2. Mixed Oxides

Mixed oxide ceramics that have been prepared by ECAS in transparent or translucent form include titanates and zirconates with perovskite structure (barium titanate $BaTiO_3$, strontium titanate $SrTiO_3$, barium strontium titanate $Ba_xSr_{(1-x)}TiO_3$, La-doped lead zirconate titanate / PLZT,

Ba-Zr-Ce-Y-oxide perovskite $BaZr_{0.5}Ce_{0.3}Y_{0.2}O_3$, lead zirconate titanate / PZT-based $PbZrO_3$-$PbTiO_3$-$Pb(Zn_{0.33}Nb_{0.66})O_3$ ceramics) and several lutetium compounds (lutetium niobate Lu_3NbO_7, lutetium titanate pyrochlore $Lu_2Ti_2O_7$, lutetium hafnium oxide $Lu_2Hf_2O_7$ and lutetium aluminum garnet $Lu_3Al_5O_{12}$ /LuAG), as well as strontium hafnium oxide $SrHfO_3$ and zinc aluminate spinel $ZnAl_2O_4$, but mainly – and most importantly – magnesium aluminate spinel $MgAl_2O_4$ and yttrium aluminum garnet $Y_3Al_5O_{12}$ /YAG, see Appendix B2.

While the titanates and zirconates are candidates for electro-optical devices, such as optical modulation devices, optical shutters and switchers, as well as image memory devices, potential applications of the lutetium compounds are phosphor and scintillation materials, ionic conductors, laser gain media, and photocatalysts (Abe et al. 2006, Brixner 1984, Duan et al. 2009, Fujihara & Tokumo 2005, Haertling & Land 1971, Shlyakhtina et al. 2005, Yanagida et al. 2009).

Yttrium aluminium garnet (YAG) is one of the most thoroughly investigated and most frequently prepared transparent ceramic materials today. Its importance consists mainly in the fact that the yttrium ion can be partially or totally replaced by rare earth element (REE) ions, because both the valency and the ion radius of Y^{3+} is close to that of REE ions. Therefore YAG is an ideal host for laser-active REE ions and thus an ideal material for solid state lasers. The cubic crystal structure of YAG and the ensuing optical isotropy of the crystallites make YAG a promising material for high transparency, as required in solid state laser applications. However, the presence of residual secondary phases from the solid state reaction of alumina and yttria (i.e., orthorhombic yttrium aluminum perovskite $YAlO_3$ / YAP or monoclinic yttrium aluminium oxide $Y_4Al_2O_9$ / YAM) and the formation of new silicon-containing secondary phases at the grain boundaries (from the tetraethyl orthosilicate / TEOS, which is often used as a sintering aid) has to be avoided. Therefore it is usually recommended to use ready-to-use YAG powder instead of a reactive sintering route based on Al_2O_3 and Y_2O_3 powders for the fabrication of transparent YAG ceramics (Ikesue et al. 2013).

In contrast to YAG, which is mainly used for solid state lasers, but also as a phosphor and scintillator material (Ikesue 2000, Ikesue et al. 1995), but similar to zinc aluminate spinel ($ZnAl_2O_4$), which is used as for thermal shock resistant high-temperature optical windows (van der Laag et al. 2004), also magnesium aluminate spinel ($MgAl_2O_4$) is mainly used for optical windows, including windows for spacecraft, UV and IR laser windows, transparent armor and optical components of space telescopes (Harris 2005).

For magnesium aluminate spinel ceramics, which have been prepared in transparent form via ECAS by different authors (Alekseev et al. 2016, Benaissa et al. 2016, Bernard-Granger et al. 2009, Biswas et al. 2016, Bonnefont et al. 2012, Boulesteix et al. 2017, Frage et al. 2007, Fu et al. 2013, 2014, 2015, Kim et al. 2010a, Meir et al. 2009, Morita et al. 2008, 2009, 2012, 2016, Nam et al. 2017, Rothmann et al. 2014, Sokol et al. 2014a, 2017, Wang & Zhao 2009), the sintering temperatures range from 900 to 1600°C, but are typically in the range 1250–1500°C, with heating rates up to 300°C/min, maximum pressures up to several hundred MPa (typically 80 MPa) and dwell times up to 120 min (typically around 20 min). For specimens with thicknesses in the range 1.0–2.5 mm the absolute transmittances in the visible region are in the range 35–83%, whereas transmittances at infrared wavelengths are up to 83% for specimens as thick as 3.0 mm. LiF as a sintering aid and post-annealing treatments (at temperatures in the range 800–1300°C for times of 1 h) have often been used to eliminate or at least reduce the yellow or dark (grey or brown) discoloration that has often been observed.

For zinc aluminate spinel ceramics, which have been prepared in transparent form via ECAS by different authors (Yong et al. 2014, Kim et al. 2014, Fu et al. 2017), the sintering temperatures range from 1200 to 1280°C, with heating rates from 10 to 100°C/min, pressures in the range 80–100 MPa and dwell times of 15-20 min. For specimens with thickness 1.0 mm the absolute transmittances in the visible region are in the range 64–77%, whereas transmittances at infrared wavelengths are up to 85–87% for specimens as thick as 1.0 mm. In one case (Yong et al. 2014) TEOS has been used as a sintering aid, and in one case (Kim et al. 2014) post-annealing (at 1100°C for 3 h) has been used.

For yttrium aluminum garnet ceramics, which have been prepared in transparent form via ECAS by different authors (Chaim et al. 2006, 2007, Frage et al. 2010, 2012, Penilla et al. 2012, Sokol et al. 2014b, Spina et al. 2012, Suárez et al. 2009, Wang et al. 2016, 2017, Zhang et al. 2016), the sintering temperatures range from 700 to 1550°C, but are typically in the range 1200–1500°C, with heating rates up to 200°C/min (typically 100°C/min), maximum pressures in the range 50–100 MPa and dwell times up to 120 min (but usually below 50 min, typically a few minutes). For specimens with thicknesses in the range 0.5–3.0 mm the absolute transmittances in the visible region are in the range 14–76%, whereas transmittances at infrared wavelengths are in the range 55–82% for specimens with thickness 0.9–2.0 mm. LiF as a sintering aid or post-annealing treatments (at temperatures in the range 900–950°C for times of 1–12 h) have sometimes been used to eliminate or at least reduce the dark (grey or brown) discoloration that has been observed in some cases.

For the aforementioned titanate and zirconate ceramics, for which ECAS has been used by various authors (Liu et al. 2010, Wu et al. 2002, 2005, 2011, Xing et al. 2017), the sintering temperatures range from 700 to 1000°C (with the exception of $BaZr_{0.5}Ce_{0.3}Y_{0.2}O_3$, for which it was 1350°C), but are typically in the range 900–1000°C, with heating rates up to 100°C/min, maximum pressures in the range 29–200 MPa (23 MPa for $BaZr_{0.5}Ce_{0.3}Y_{0.2}O_3$) and dwell times of 3–10 min. For specimens with thicknesses in the range 0.3–0.9 mm the absolute transmittances in the visible region are in the very wide range 4–74%, whereas transmittances at infrared wavelengths are in the range 50–94% for specimens with thickness 0.8–0.9 mm. Post-annealing treatments have always been performed (at temperatures in the range 700–1200°C for times of 1–3 h), but LiF as a sintering aid has usually not been used. It is evident that the transparency of zirconates and titanates prepared by ECAS is usually not very good. A similar conclusion holds for strontium hafnium oxide $SrHfO_3$ prepared via ECAS at 1700°C with heating rates of 17–200°C/min, pressure of 100 MPa and 45 min dwell. Despite the harsh conditions, the transmittance is only 10% at a wavelength of 640 nm (Kurosawa et al. 2013).

Finally, for the aforementioned lutetium compounds, to which ECAS has been applied by An and coworkers (An et al. 2011a, 2011e, 2011f, 2012b, 2012c, 2013a, 2013b), the sintering temperatures range from 1000 to 1650°C, with heating rates from 10 to 200°C/min, pressures of 100 MPa and dwell times of 45 min. For specimens with thickness 1.0 mm the absolute transmittances in the visible region are in the range 34–68%, whereas transmittances at infrared wavelengths are in the range 71–78%. Post-annealing treatments have always been performed (at temperatures in the range 650–1550°C for times of 6–60 h), but LiF as a sintering aid has not been used.

5.3. Phosphates and Silicates

Also phosphate and silicate ceramics have been prepared by ECAS in transparent or translucent form. Hydrodroxyapatite/$Ca_{10}(PO_4)_6(OH)_2$, often abbreviated as HA, is a popular material for bone tissue engineering and biomedical implant applications, due to its biocompatibility and bioactivity (Dorozhkin 2010). Its use in transparent form may be a viable route to use doping with appropriate ions in order to cause specific absorption and / or re-emission of light or other electromagnetic radiation of certain wavelengths and thus lead to local heating and radiation in vivo, an effect that might be of future use in cancer therapy, for example. Although similar conclusions may hold for tricalcium phosphate/$Ca_3(PO_4)_2$, often abbreviated as TCP, the formation of TCP during the sintering of HA is usually undesirable. Actually, a fundamental problem in the conventional sintering of HA ceramics is the fact that in order to densify HA the temperature should be sufficiently high, but in order to avoid TCP formation the temperature must not be too high. Therefore, HA ceramics made by conventional sintering are usually porous and too weak to be used directly as load-bearing implants. This is the reason why HA materials are commonly used either in the form of spherical granules as a bone filler material or as a porous scaffold for bone tissue engineering, where the osteoconductive and osteoinductive scaffold is inoculated with bone stem cells and growth factors to stimulate ingrowth of differentiated bone cells. Only after this tissue engineering step the hybrid HA biocomposite would be usable as an implant. ECAS, on the other hand, may offer the possibility to obtain dense and this mechanically relatively strong HA ceramics without the formation of TCP.

A rather exotic material is strontium fluorapatite/$Sr_5(PO_4)_3F$, which has been proposed as a host material for solid-state lasers (Chen et al. 2013). Similar applications (laser host materials and scintillator materials) are intended for a complex silicate with apatite structure, so-called strontium yttrium oxyapatite/$Sr_2Y_8(SiO_4)_6O_2$ (Druon et al. 2002). Another silicate with apatite structure is lanthanum oxyapatite/$La_{9.33}Si_6O_{26}$, which has been proposed as an electrolyte in solid oxide fuel cells/SOFCs (Nakayama & Sakamoto 1998). All these materials have been prepared by ECAS.

Among the more common silicates it seems that only mullite/$Al_6Si_2O_{13}$ has been prepared as a single-phase transparent ceramic via ECAS so far. Mullite is the most common crystalline phase in many traditional silicate ceramics, such as porcelain, and an interesting ceramic material in its own right (Schneider & Komarneni 2005). Due to its excellent mechanical properties, refractoriness and corrosion resistance it is suited for many advanced structural applications. Although mullite is non-cubic, its birefringence is relatively small (0.015), compared to 0.007 for hydroxyapatite, 0.008 for alumina (Bernard & Rost 1992) and 0.030–0.040 for tetragonal zirconia, which has also been prepared in transparent form (Klimke et al. 2011). Therefore, transparent mullite ceramics is a promising candidate material for optical windows in high-temperature applications (Aksay 1991, Schneider 2008).

For hydroxyapatite ceramics, which have been prepared in transparent form via ECAS by different authors (Chaudhry et al. 2011, Eriksson et al. 2011, Han 2016, Ioku et al. 2002, Kim et al. 2013, Li et al. 2016b, Watanabe et al. 2005, Yun et al. 2014), see Appendix B3, the sintering temperatures range from 700 to 1200°C, but are typically in the range 900–1000°C, with heating rates ranging from 20 to 200°C/min, maximum pressures in the range 10–500 MPa (typically 50–100 MPa) and dwell times of a few minutes (3–10 min). For specimens with thicknesses in the range 0.7–2.7 mm the absolute transmittances in the visible region are in the range 45–70%, whereas at infrared wavelengths a transmittance of 84% has been reported for a specimen with thickness 2.0 mm. Sintering aids or post-annealing treatments have not been used. Transparent tricalcium phosphate (TCP) ceramics have been prepared by Kawagoe and coworkers (Kawagoe et al. 2004) at sintering temperatures of 800–1000°C with heating rate 25°C and pressure 60 MPa, without any dwell, sintering aids or post annealing. Although transparency has been reported, transmittance values have not been given in this case.

Neither the lanthanum oxyapatite/$La_{9.33}Si_6O_{26}$ ceramics, prepared (without doping or sintering aids) via ECAS by Chesnaud and coworkers (Chesnaud et al. 2007) at 1200–1500°C with heating rates of 50–300°C/min, a pressure of 100 MPa, a dwell time of 3 min and a post-annealing step (12 h at 1000°C), nor the strontium yttrium oxyapatite/$Sr_2Y_8(SiO_4)_6O_2$, prepared via ECAS by Shen and coworkers (Shen et al. 2010) at 1400–1500°C with heating rates of 100–300°C/min, a pressure of 23 MPa, a dwell time of 3 min and a post-annealing step (4 h at 1400°C), were really transparent. At best, a certain degree of translucency could be achieved. This seems to indicate a fundamental problem with these silicates. By contrast, the 20 mol% Yb-doped strontium fluorapatite/$Sr_5(PO_4)_3F$, prepared via ECAS by Chen and coworkers (Chen et al. 2013) at 1050°C with a heating rate of 50°C/min, a pressure of 100 MPa, a dwell time of 8 min and a post-annealing step (4 h at 1400°C), exhibited transmittances of 50% and 74% at wavelengths 500 nm and 1000 nm, respectively.

For mullite ceramics, which have been prepared in transparent form via ECAS by different authors either from commercial mullite powder (Kocjan et al. 2016) or from powder specially prepared via sol-gel processes (Ren et al. 2015, Zhang et al. 2009), see Appendix B3, the sintering temperatures that have been tested range from 1250 to 1600°C, using heating rates of 100°C/min, maximum pressures in the range 15–80 MPa and dwell times in the range 2–20 min). For a specimen with thickness 1.0 mm the absolute transmittance achieved in the visible region (640 nm) was rather low (20%), but for other specimens with thicknesses in the range 0.6–1.0 mm the transmittance achieved in the infrared region (2000–4300 nm) was quite high (75–83%), which seems to predetermine transparent mullite ceramic windows for applications in infrared detector systems. Neither sintering aids nor post-annealing have been used for preparing transparent mullite ceramics via ECAS.

5.4. Non-Oxides, Oxynitrides and Composites

Non-oxide ceramics that can be prepared in transparent form include nitrides and oxynitrides, such as aluminum nitride (AlN), silicon nitride (Si_3N_4), AlON and SiAlON, but also some fluorides of alkaline earths, such as calcium fluoride (CaF_2) and magnesium fluoride (MgF_2), as well as special materials, such as sphalerite (ZnS), see (Siqueiros et al. 1988). Actually, the latter has been used as a material for infrared windows and missile domes in defense applications for a very long time (Greene 1958). Also calcium and magnesium fluoride, which are insoluble in water (in contrast to alkali halides), are traditional materials for optical components (windows and lenses) in spectroscopic equipment, space telescopes, excimer lasers and thermal imaging systems. Due to their wide range of transparency from the ultraviolet to the infrared part of the electromagnetic spectrum they are ideally suited for these applications. More recently, they have been proposed as materials for sub-100fs laser oscillators (Friebel et al. 2009, Jounini et al. 2008, Lucca et al. 2004). Aluminum nitride, silicon nitride and their oxynitride counterparts (AlON and SiAlON) are promising materials for optical windows that are mechanically strong, hard and resistant against scratching, abrasion and impact, i.e., ideally suited for transparent armor applications (Harris 1998, Yang et al. 2013, Swab et al. 2014).

For calcium and magnesium fluoride ceramics, which have been prepared in transparent form via ECAS by different authors (Li et al. 2016a, Nakamura et al. 2017a, 2017b, 2017c, 2017d, 2018), see Appendix

B4, the sintering temperatures that have been tested are in the range 600–1100°C for CaF_2 and 650–750°C magnesium fluoride MgF_2, with heating rates from 10 to 100°C/min (typically 80–100°C/min), pressures in the range 30–100 MPa (typically 70 MPa) and dwell times of 5–15 min. For specimens with thicknesses 0.5–2.0 mm the absolute transmittances in the visible region are in the range 12–61%, whereas transmittances at infrared wavelengths are up to 83–86% for specimens with thicknesses 1.0–2.0 mm. Doping with REE ions is frequent, but sintering aids and post-annealing have not been used. For sphalerite ceramics, prepared (without doping, sintering aids or post-annealing) via ECAS by Chen and coworkers (Chen et al. 2015) at 850°C with heating rates up to 100°C/min, a pressure of 80 MPa and a dwell time of 5 min a transmittance of 68% has been achieved in the IR region (at wavelength 5000 nm) for a specimen with thickness 1.5 mm.

For aluminum nitride, prepared in transparent form via ECAS (Fu et al. 2004, Xiong et al. 2005) by sintering at 1800°C (with or without 0–3 wt% of fluorite/CaF_2 as a sintering aid) with heating rates of 100°C/min (although higher heating rates up to 550°C/min have been tested as well), pressures in the range 25–30 MPa and dwell times of 4–20 min, a transmittance of 55% has been achieved for a 0.4 mm thick specimen (prepared with 3 wt% of CaF_2) at infrared wavelengths (5500 nm). In this case the AlN powder was ball-milled with the sintering aid and ECAS was carried out in nitrogen atmosphere (Xiong et al. 2005).

Similarly, for aluminum oxynitride (AlON), prepared in transparent or translucent form via ECAS (Kaplan et al. 2015, Shan et al. 2017) by sintering in the range 1350–1700°C (with different dopants or sintering aids, see Appendix B4) with heating rates of 30–100°C/min, pressures in the range 30–40 MPa and dwell times of 1–30 min, transmittances of up to 77% have been achieved for 1.2 mm thick specimens at infrared wavelengths (3900 nm). After ECAS these transparent specimens, prepared at a relatively low temperature (1350–1500°C) with 0.5 wt% Y_2O_3 as a sintering aid (heating rate 100°C/min, pressure 40 MPa, dwell time 10 min), contained three phases, viz. α-AlON, γ-Al_2O_3 and h-AlN (Shan et al. 2017). On the other hand, specimens prepared with Dy^{3+} and

Mg^{2+} doping at higher temperature (1500–1700°C) with a slower heating rate (30°C/min), similar pressure (30 MPa) and variable dwell times (1–30 min) attained only transmittances of 18% at 5680 nm (for unspecified specimen thickness), although a (non-specified) post-annealing treatment has been performed in an attempt to remove residual Al_2O_3 and AlN (Kaplan et al. 2015).

The results for aluminum nitride and oxynitride may be compared with the results for translucent silicon aluminum oxynitride (α-SiAlON) ceramics, prepared from ball-milled mixtures of commercial Si_3N_4, AlN and MgO powders via ECAS (Xiong et al. 2007, Yang et al. 2013) by sintering at 1850°C (with 3 wt% MgO as a sintering aid) with a heating rate of 100°C/min, a pressure of 30 MPa and a dwell time of 5 min. Transmittances of up to 66% have been achieved for 0.2–0.5 mm thick specimens at infrared wavelengths (5470–5500 nm). Although grey discoloration has been commonly observed, post-annealing has not been used for nitrides and oxynitrides. The poor transmittance in the IR region of β-S_3N_4 has been attributed to nitrogen vacancies (Yang et al. 2013).

Transparent composite ceramics prepared via ECAS belong to a very special class of materials. It is clear that composites are multiphase materials, for which usually strong light attenuation due to scattering must be expected. On the other hand, when either the refractive indices match or the scattering centers are small enough it is not impossible to obtain transparent composites. In cases where this can be achieved it is obvious that additional benefits can be obtained by combining appropriate optical properties with other favorable features of behavior, e.g., higher strength, hardness, fracture toughness or, for that matter, thermal conductivity.

There are at least two published works dealing with transparent yttria-magnesia nanocomposite ceramics (Jiang & Mukherjee 2010, Xu et al. 2015). In the first one (Jiang & Mukherjee 2010) the Y_2O_3-MgO composite powder in the mass ratio of 1:1 was prepared by flame pyrolysis, followed by heat treatment at 1100°C in air for 30 min and ultrasonication. This procedure resulted in a fluffy powder with a nanoscale particle size of approximately 60 nm. ECAS was carried out at temperatures in the range 1200–1600°C with heating rates in the range 80–200°C/min, pressure 80

MPa and dwell times of 3–8 min, followed by post-annealing at 1000°C in air for 15 h. The best in-line transmittance obtained was 85% at an infrared wavelength of 4000 nm, obtained for a specimen of thickness 1.0 mm, sintered at 1200°C with a dwell time of 3 min, resulting in a grain size of approximately 90 nm in the material. Similarly, in the second work (Xu et al. 2015) an infrared-transparent Y_2O_3-MgO nanocomposite was fabricated using powder synthetized by a sol-gel method. Calcination was carried out at 800°C in air for 1 h and resulted in nanoscale particles with a size of approximately 13 nm. ECAS was carried out at 1000–1300°C with a heating rate of 100°C/min, maximum pressure of 560 MPa and a dwell time of 5 min. The dark discoloration that appeared in the specimens after ECAS faded out after post-annealing at 1000°C in air for 15 h. The best in-line transmittance obtained was 58% at an infrared wavelength of 4000 nm, obtained for a specimen of thickness 1.2 mm, sintered at 1100°C, resulting in a final grain size of approximately 100 nm.

CONCLUSION

In this chapter the application of electric current assisted sintering (ECAS) for the preparation of transparent ceramics has been reviewed. The principles of the ECAS technique have been explained and, based on classical scattering theory (Mie theory), a concise overview has been given on the fundamental microstructure-property relations that determine transparency (transmittance), mainly the influence of the concentration (volume fraction), size (equivalent diameter) and optical properties of the inclusions. It has been shown that ECAS is a sintering technique that can be used for the fabrication of very dense ceramics with a certain and sometimes remarkable degree of transparency. However, often this transparency is impaired by discoloration, mainly due to oxygen vacancies and / or carbon contamination.

As far as specific ECAS conditions and schedules for individual ceramic materials are concerned, this chapter has shown that the maximum sintering temperatures tested and used by different authors in the context of transparent ceramics range from 600°C to 1800°C, but are mostly in the range 1000–1400°C, of course very much depending on the material. The maximum heating rates used can reach several hundred °C/min (up to 1000°C/min), but are usually around 100°C/min (often varying within a single ECAS schedule). Similarly, the pressures are usually limited to approximately 100 MPa when a simple graphite die is used, although much higher pressure can be used with special compound dies (consisting of an inner and outer graphite die) or with cemented carbide dies (WC, SiC). Commonly used dwell times range from a few minutes to about one hour, although much longer dwell have been tested in singular cases. In general, it may be said that, similar to conventional hot pressing techniques, the increased pressure supports densification at lower temperature and thus suppresses grain growth, while the high heating rates (and possibly short dwell times) make ECAS an economically interesting option for ceramics, with which more conventional hot pressing processes cannot compete. The resulting materials are very dense and mechanically strong. The ceramic material classes to which ECAS has been successfully applied and for which good transparency has been attained, include simple oxides (mainly alumina, lutecia, magnesia, yttria and zirconia), mixed oxides (mainly titanates and zirconates, several lutetium compounds, magnesium aluminate spinel and yttrium aluminium garnet), phosphates and silicates (mainly hydroxyapatite, complex silicates with apatite structure and mullite) as well as non-oxides (mainly aluminum nitride and oxynitride and silicon aluminum oxynitride) and even composites (yttria-magnesia nanocomposites).

As far as the optical properties are concerned, transparent ceramics prepared via ECAS often exhibit discoloration. This discoloration has been attributed by different authors either to oxygen vacancies (47.4% of the papers cited), carbon contamination (28.2%) or both effects (20.5%). Other causes are invoked only very rarely (3.9%), i.e., in exceptional cases (e.g., as a consequence of sulphur or fluorine impurities). In 50% of all cases (optimum results) the relative transmittance achieved was 0.72 or higher, i.e., at least 72% of the theoretical maximum transmittance (which is dictated by the refractive index of a given solid material). In approximately

36% of the papers cited relative transmittances higher than 0.90, i.e., more than 90% of the theoretical maximum transmittance, have been attained. These values may serve as rough guidelines for defining realistic material design targets with ECAS. However, it has to be recalled that the relative transmittance depends on the specimen thickness and that the relative transmittance of different materials can only be directly compared when the specimen thickness is the same. Therefore it is useful to calculate the extinction coefficients, which are independent of the specimen thickness (at least when the concentration of scattering centers is low, so that multiple scattering is excluded) and can under certain conditions (high-phase-contrast inclusions very small or large compared to the wavelength of light) be considered as relative material properties that directly reflect the effect of the microstructure, i.e., the concentration and size of the inclusions, on the relative transmittance. The statistical evaluation of the literature data collected in this review (taking into account only the most successful, optimum results of each work) yielded a median value of the extinction coefficient of around 0.3, i.e., in 50% of all cases the extinction coefficient was lower (down to 0.002) and in 50% of all cases the extinction coefficient was higher (up to 4.1). Unfortunately, current literature data do not allow correlating the individual extinction coefficients directly with the individual microstructures, because in the absence of a convenient method for their measurement it is not surprising that the pore sizes or sizes of carbon inclusions are not adequately characterized. Nevertheless, these values may serve as a certain benchmark for future research and as first estimates when input information is needed for materials design. The cumulative distribution of the extinction coefficients gives a rough "material-independent" impression of the relative transparency that can realistically be achieved via ECAS. Of course, when a specific material is to be used, the reader has to refer to the original literature for that specific material and the ECAS conditions and schedules that have been investigated for that material so far. For this purpose the present chapter provides handy guidelines, on which further research and development can be based.

Acknowledgments

This work is part of the project "Partially and fully sintered ceramics – processing, microstructure, properties, modeling and sintering theory" (GA18-17899S), supported by the Czech Science Foundation (GAČR). The authors would like to thank their colleagues Soňa Hříbalová for performing the Mie scattering calculations and Tereza Uhlířová for efficient help with processing the literature references.

References

Abe, R.; Higashi, M.; Sayama, K.; Abe, Y.; Sugihara, H. *J. Phys. Chem. B* 2006, *110,* 2219-2226.

Ahmadi, B.; Reza, S. R.; Ahsanzadeh-Vadeqani, M.; Barekat, M. *Ceram. Int.* 2016, *42,* 17081-17088.

Ahsanzadeh-Vadeqani, M.; Shoja Razavi, R. *Ceram. Int.* 2016, *42,* 18931-18936.

Aksay, I. A.; Dabbs, D. M.; Sarikaya, M. *J. Am. Ceram. Soc.* 1991, *74,* 2343-2358.

Alaniz, J. E.; Perez-Gutierrez, F. G.; Aguilar, G.; Garay, J. E. *Opt. Mater.* 2009, *32,* 62-68.

Alekseev, M. K.; Kulikova, G. I.; Rusin, M. Y.; Savanina, N. N.; Balabanov, S. S.; Belyaev, A. V.; Gavrishchuk, E. M.; Ivanov, A. V.; Rizakhanov, R. N. *Inorg. Mater.* 2016, *52,* 324-330.

Alombert-Goget, G.; Guyot, Y.; Guzik, M.; Boulon, G.; Ito, A.; Goto, T.; Yoshikawa, A.; Kikuchi, M. *Opt. Mater.* 2015, *41,* 3-11.

Alvarez-Clemares, I.; Mata-Osoro, G.; Fernández, A.; Lopez-Esteban, S.; Pecharromán, C.; Palomares, J.; Torrecillas, R.; Moya, J. S. *Adv. Eng. Mater.* 2010, *12,* 1154-1160.

Alvarez-Clemares, I.; Mata-Osoro, G.; Fernández, A.; Lopez-Esteban, S.; Pecharromán, C.; Torrecillas, R.; Moya, J. S. *J. Eur. Ceram. Soc.* 2012, *32,* 2917-2924.

Aman, Y.; Garnier, V.; Djurado, E. *J. Eur. Ceram. Soc.* 2009, *29,* 3363-3370.

Aman, Y.; Garnier, V.; Djurado, E. *Int. J. Appl. Ceram. Technol.* 2010, *7,* 574-586.

An, L.; Ito, A.; Goto, T. *J. Eur. Ceram. Soc.* 2011, *31,* 237-240.

An, L.; Ito, A.; Goto, T. *J. Am. Ceram. Soc.* 2011, *94,* 695-698.

An, L.; Ito, A.; Goto, T. *J. Eur. Ceram. Soc.* 2011, *31,* 1597-1602.

An, L.; Ito, A.; Goto, T. *Ceram. Int.* 2011, *37,* 2263-2267.

An, L.; Ito, A.; Goto, T. *J. Am. Ceram. Soc.* 2011, *94,* 3851-3855.

An, L.; Ito, A.; Goto, T. *Mater. Lett.* 2011, *65,* 3167-3169.

An, L.; Ito, A.; Goto, T. *J. Eur. Ceram. Soc.* 2012, *32,* 1035-1040.

An, L.; Ito, A.; Goto, T. *Ceram. Int.* 2012, *38,* 4973-4977.

An, L.; Ito, A.; Goto, T. *J. Eur. Ceram. Soc.* 2012, *32,* 3097-3102.

An, L.; Ito, A.; Goto, T. *Ceram. Int.* 2013, *39,* 383-387.

An, L.; Ito, A.; Goto, T. *Opt. Mater.* 2013, *35,* 817-819.

An, L.; Ito, A.; Goto, T. *J. Asian Ceram. Soc.* 2014, *2,* 154-157.

An, L.; Ito, A.; Zhang, J.; Tang, D.; Goto, T. *Opt. Mater. Express* 2014, *4,* 1420-1426.

Apetz, R.; van Bruggen, M. P. B. *J. Am. Ceram. Soc.* 2003, *86,* 480-486.

Attia, M. A. A.; Orrù, R.; Delogu, F.; Montinaro, S.; Garroni, S.; Ewais, E. M. M.; Cao, G.; Kleebe, H. J. *J. Am. Ceram. Soc.* 2015, *98,* 1453-1459.

Benaissa, S.; Hamidouche, M.; Kolli, M.; Bonnefont, G.; Fantozzi, G. *Ceram. Int.* 2016, *42,* 8839-8846.

Bernard, J. H.; Rost, R. *Encyclopedický přehled minerálů [Encyclopedic Overview on Minerals];* Academia: Prague, Czechoslovakia, 1992, pp 243-301.

Bernard-Granger, G.; Benameur, N.; Guizard, C.; Nygren, M. *Scr. Mater.* 2009, *60,* 164-167.

Biswas, P.; Chakravarty, D.; Suresh, M. B.; Johnson, R.; Mohan, M. K. *Ceram. Int.* 2016, *42,* 17920-17923.

Bloxam, A. G. GB Patent No. 27,002 (1906).

Bloxam, A. G. GB Patent No. 9020 (1906).

Bohren, C. F.; Huffman, D. R. *Absorption and Scattering of Light by Small Particles;* Wiley: New York, NY, 1983 (reprint Wiley-VCH: Weinheim, Germany, 2004), pp 12-165.

Bond, W. L. *J. Appl. Phys.* 1965, *36,* 1674-1677.

Bonnefont, G.; Fantozzi, G.; Trombert, S.; Bonneau, L. *Ceram. Int.* 2012, *38,* 131-140.

Boulesteix, R.; Epherre, R.; Noyau, S.; Vandenhende, M.; Maître, A.; Sallé, C.; Alombert-Goget, G.; Guyot, Y.; Brenier, A. *Scripta Mater.* 2014, *75,* 54-57.

Boulesteix, R.; Maître, A.; Lemański, K.; Dereń, P. J. *J. Alloys Comp.* 2017, *722,* 358-364.

Brixner, L. H. *Mater. Res. Bull.* 1984, *19,* 143-149.

Burke, J. E. The development of pore-free ceramics, in *Sintering Key Papers;* Somiya, S.; Moriyoshi, Y. (Eds.); Elsevier Applied Science: London, UK, 1990; pp 3-16.

Casolco, S. R.; Xu, J.; Garay, J. E. *Scripta Mater.* 2008, *58,* 516-519.

Chaim, R.; Shen, Z.; Nygren, M. *J. Mater. Res.* 2004, *19,* 2527-2531.

Chaim, R.; Marder-Jaeckel, R.; Shen, J. Z. *Mater. Sci. Eng. A* 2006, *429,* 74-78.

Chaim, R.; Kalina, M.; Shen, J. Z. *J. Eur. Ceram. Soc.* 2007, *27,* 3331-3337.

Chaudhry, A. A.; Yan, H.; Gong, K.; Inam, F.; Viola, G.; Reece, M. J.; Goodall, J. B.; ur Rehman, I.; McNeil-Watson, F. K.; Corbett, J. C.; Knowles, J. C.; Darr, J. A. *Acta Biomater.* 2011, *7,* 791-799.

Chen, S.; Wu, Y.; Yang, Y.; Bhandarkar, S. *J. Am. Ceram. Soc.* 2013, *96,* 1694-1697.

Chen, Y.; Zhang, L.; Zhang, J.; Liu, P.; Zhou, T.; Zhang, H.; Gong, D.; Tang, D.; Shen, D. *Opt. Mater.* 2015, *50,* 36-39.

Chesnaud, A.; Bogicevic, C.; Karolak, F.; Estournes, C.; Dezanneau, G. *Chem. Commun.* 2007, *2007,* 1550-1552.

Cremer, G. D. US Patent No. 2,355,954 (1944).

Debenham, M. *Appl. Opt.* 1984, *23,* 2238-2239.

Dodge, M. J. *Appl. Opt.* 1984, *23,* 1980-1985.

Dorozhkin, S. V. *Biomaterials* 2010, *31,* 1465-1485.

Druon, F.; Chévais, S.; Raybaut, P.; Balembois, F.; Georges, P.; Gaumé, R.; Haumesser, P. H.; Viana, B.; Vivien, D.; Dhellemmes, S.; Ortiz, V.; Larat, C. *Opt. Lett.* 2002, *27,* 1914-1916.

Du, P.; Huo, D.; Xi, H.; Kang, Y.; Sun, X.; Xiao, D. *Ceram. Int.* 2017, *43,* 9854-9859.

Duan, X.; Yao, B.; Li, G.; Ju, Y.; Wang, Y.; Zhao, G. *Opt. Express* 2009, *17,* 21691-21697.

Duval d'Adrian, A. L. US Patent No. 1,430,724 (1922).

Eriksson, M.; Liu, Y.; Hu, J.; Gao, L.; Nygren, M.; Shen, Z. *J. Eur. Ceram. Soc.* 2011, *31,* 1533-1540.

Esposito, L.; Hostaša, J.; Piancastelli, A.; Toci, G.; Alderighi, D.; Vannini, M.; Epicier, T.; Malchère, A.; Alombert-Goget, G.; Boulon, G. *J. Mater. Chem. C* 2014, *2,* 10138-10148.

Frage, N.; Cohen, S.; Meir, S.; Kalabukhov, S.; Dariel, M. P. *J. Mater. Sci.* 2007, *42,* 3273-3275.

Frage, N.; Kalabukhov, S.; Sverdlov, N.; Ezersky, V.; Dariel, M. P. *J. Eur. Ceram. Soc.* 2010, *30,* 3331-3337.

Frage, N.; Kalabukhov, S.; Sverdlov, N.; Kasiyan, V.; Rothman, A.; Dariel, M. P. *Ceram. Int.* 2012, *38,* 5513-5519.

Frei, J. M.; Anselmi-Tamburini, U.; Munir, Z. A. *J. Appl. Phys.* 2007, *101,* 114914 (8 pp).

Friebel, F.; Druon, F.; Boudeile, J.; Papadopoulos, D. N.; Hanna, M.; Georges, P.; Camy, P.; Doualan, J. L.; Benayad, A.; Moncorgé, R.; Cassagne, C.; Boudebs, G. *Optics Lett.* 2009, *34,* 1474-1476.

Fu, Z. Y.; Liu, J. F.; Wang, H.; He, D. H.; Zhang, Q. J. *Mater. Sci. Technol.* 2004, *20,* 1097-1099.

Fu, P.; Lu, W.; Lei, W.; Xu, Y.; Wang, X.; Wu, J. *Ceram. Int.* 2013, *39,* 2481-2487.

Fu, P.; Xu, Y.; Shi, H.; Zhang, B.; Ruan, X.; Lu, W. *Opt. Mater.* 2014, *36,* 1232-1237.

Fu, P.; Xu, Y.; Lu, W.; Lei, W.; Wang, X.; Ruan, X. *Int. J. Appl. Ceram. Technol.* 2015, *12,* 116-123.

Fu, P.; Wang, Z.; Lin, Z.; Liu, Y.; Roy, V. A. L. *J. Mater. Sci.: Mater Electron.* 2017, *28,* 9589-9595.

Fujihara, S.; Tokumo, K. *Chem. Mater.* 2005, *17,* 5587-5593.

Furuse, H.; Nakasawa, S.; Yoshida, H.; Morita, K.; Suzuki, T. S.; Kim, B. N.; Sakka, Y.; Hiraga, K. *J. Am. Ceram. Soc.* 2018, *101,* 694-702.

Garay, J. E. *Ann. Rev. Mater. Res.* 2010, *40,* 445-468.

Ghaderi, M.; Shoja Razavi, R.; Loghman-Estarki, M. R.; Ghorbani, S. *Ceram. Int.* 2016, *42,* 14403-14410.

Ghanizadeh, S.; Grasso, S.; Ramanujam, P.; Vaidhyanathan, B.; Binner, J.; Brown, P.; Goldwasser, J. *Ceram. Int.* 2017, *43,* 275-281.

Gorokhova, E. I.; Ananeva, G. V.; Domidenko, V. A.; Rodny, P. A.; Khodyuk, I. V.; Bourret-Courchesne, E. D. *J. Opt. Technol. (Opticheskii Zhurnal)* 2008, *75,* 741-746.

Grasso, S.; Sakka, Y.; Maizza, G. *Sci. Technol. Adv. Mater.* 2009, *10,* 053001 (24 pp).

Grasso, S.; Kim, B. N.; Hu, C.; Maizza, G.; Sakka, Y. *J. Am. Ceram. Soc.* 2010, *93,* 2460-2462.

Grasso, S.; Hu, C.; Maizza, G.; Kim, B.-N.; Sakka, Y. *J. Am. Ceram. Soc.* 2011, *94,* 1405-1409.

Greene, L. C.; Reynolds, D. C.; Czyzak, S. J.; Baker, W. M. *J. Chem. Phys.* 1958, *29,* 1375-1380.

Guillon, O.; Gonzalez-Julian, J.; Dargatz, B.; Kessel, T.; Schierning, G.; Räthel, J.; Herrmann, M. *Adv. Eng. Mater.* 2014, *16,* 830-849.

Han, Y. H.; Kim, B. N.; Yoshida, H.; Yun, J.; Son, H. W.; Lee, J.; Kim, S. *Adv. Appl. Ceram.* 2016, *115,* 174-184.

Haertling, G. H.; Land, C. E. *J. Am. Ceram. Soc.* 1971, *54,* 1-11.

Harris, D. C. *Infrared Phys. Technol.* 1998, *39,* 185-201.

Harris, D. C. History of development of polycrystalline optical spinel in the US, in *Window and Dome Technologies and Materials IX;* Tustison, R. W. (Ed.); SPIE: Bellingham, WA, 2005, pp 1-22.

Hartnett, T. M.; Bernstein, S. D.; Maguire, E. A.; Tutison, R. W. *Infrared Phys. Technol.* 1998, *39,* 203-211.

Hostaša, J.; Esposito, L.; Piancastelli, A. *J. Eur. Ceram. Soc.* 2012, *32,* 2949-2956.

Hostaša, J.; Esposito, L.; Malchère, A.; Epicier, T.; Pirri, A.; Vannini, M.; Toci, G. *J. Mater. Res.* 2014, *29,* 2288-2296.

Hostaša, J.; Matějíček, J.; Nait-Ali, B.; Smith, D. S.; Pabst, W.; Esposito, L. *J. Am. Ceram. Soc.* 2014, *97,* 2602-2606.

Hu, Z.; Xu, X.; Wang, J.; Liu, P.; Li, D.; Wang, X.; An, L.; Zhang, J.; Xu, J.; Tang, D. *Ceram. Int.* 2017, *43,* 12057-12060.

Hulbert, D. M.; Anders, A.; Dudina, D. V.; Andersson, J.; Jiang, D.; Unuvar, C.; Anselmi-Tamburini, U.; Lavernia, E. J.; Mukherjee, A. K. *J. Appl. Phys.* 2008, *104,* 033305 (7 pp).

Ikesue, A. *J. Ceram. Soc. Jap.* 2000, *108,* 1020-1023.

Ikesue, A.; Aung, Y. L. *Nat. Photonics* 2008, *2,* 721-727.

Ikesue, A.; Kinoshita, T.; Kamata, K.; Yoshida, K. *J. Am. Ceram. Soc.* 1995, *78,* 1033-1040.

Ikesue, A.; Aung, Y. L.; Lupei, V. *Ceramic Lasers;* Cambridge University Press: Cambridge, UK, 2013; pp 1-445.

Inoue, K. US Patent No. 3,241,956 (1966).

Inoue, K. US Patent No. 3,250,892 (1966).

Inoue, K. US Patent No. 3,317,705 (1967).

Inoue, K. US Patent No. 3,340,052 (1967).

Ioku, K.; Kawagoe, D.; Toya, H.; Fujimori, H.; Goto, S.; Ishida, K.; Mikuni, A.; Mae, H. *Trans. Mat. Res. Soc. Japan* 2002, *27,* 447-449.

Irom, E.; Zakeri, M.; Ahangari Hossein Zadeh, A. S.; Safian, S.; Rahbari, A. *Micro Nano Lett.* 2016, *11,* 688-691.

Irom, E.; Zakeri, M.; Hossein Zadeh, A. S. A. *Ceram. Int.* 2018, *44,* 1668-1674.

Ito, A.; An, L.; Goto, T. *J. Ceram. Soc. Jap.* 2016, *124,* 313-320.

Jiang, D.; Hulbert, D. M.; Anselmi-Tamburini, U.; Ng, T.; Land, D.; Mukherjee, A. K. *J. Am. Ceram. Soc.* 2008, *91,* 151-154.

Jiang, D.; Mukherjee, A. K. *J. Am. Ceram. Soc.* 2010, *93,* 769-773.

Jiang, N.; Xie, R. J.; Liu, Q.; Li, J. *J. Eur. Ceram. Soc.* 2017, *37,* 4947-4953.

Jin, X.; Gao, L.; Sun, J. *J. Am. Ceram. Soc.* 2010, *93,* 1232-1236.

Johnson, D. L. *J. Am. Ceram. Soc.* 1990, *73,* 2576-2578.

Jouini, A.; Brenier, A.; Guyot, Y.; Boulon, G.; Sato, H.; Yoshikawa, A.; Fukuda, K.; Fukuda, T. *Cryst. Growth Des.* 2008, *8,* 808-811.

Kaminskii, A. A.; Akchurin, M. S.; Becker, P.; Ueda, K.; Bohatý, L.; Shirakawa, A.; Tokurakawa, M.; Takaichi, K.; Yagi, H.; Dong, J.; Yanagitani, T. *Laser Phys. Lett.* 2008, *5,* 300-303.

Kaplan, S. S.; Kurama, S.; Gunkaya, G. *J. Eur. Ceram. Soc.* 2015, *35,* 3255-3262.

Kato, T.; Okada, G.; Yanagida, T. *Opt. Mater.* 2016, *54,* 134-138.

Kato, T.; Okada, G.; Yanagida, T. *Radiat. Meas.* 2016, *92,* 93-98.

Kato, T.; Okada, G.; Yanagida, T. *J. Ceram. Soc. Jap.* 2016, *124,* 559-563.

Kato, T.; Okada, G.; Yanagida, T. *J. Mater. Sci.: Mater. Electron.* 2017, *28,* 7018-7023.

Kato, T.; Okada, G.; Kawaguchi, N.; Yanagida, T. *J. Lumin.* 2017, *192,* 316-320.

Kawagoe, D.; Ioku, K.; Fujimori, H.; Goto, S. *J. Ceram. Soc. Jap.* 2004, *112,* 462-463.

Kawano, N.; Kato, T.; Okada, G.; Kawaguchi, N.; Yanagida, T. *Opt. Mater.* 2017, *73,* 364-370.

Kijko, V. S.; Maksimov, R. N.; Shitov, V. A.; Demakov, S. L.; Yurovskikh, A. S. *J. Alloys Comp.* 2015, *643,* 207-211.

Kim, B.-N.; Hiraga, K.; Morita, K.; Yoshida, H. *Scripta Mater.* 2007, *57,* 607-610.

Kim, B.-N.; Hiraga, K.; Morita, K.; Yoshida, H.; Miyazaki, T.; Kagawa, Y. *Acta Mater.* 2009, *57,* 1319-1326.

Kim, B. N.; Hiraga, K.; Morita, K.; Yoshida, H. *J. Eur. Ceram. Soc.* 2009, *29,* 323-327.

Kim, B. N.; Morita, K.; Lim, J. H.; Hiraga, K.; Yoshida, H. *J. Am. Ceram. Soc.* 2010, *93,* 2158-2160.

Kim, B. N.; Hiraga, K.; Morita, K.; Yoshida, H.; Kagawa, Y. *Acta Mater.* 2010, *58,* 4527-4535.

Kim, B. N.; Hiraga, K.; Grasso, S.; Morita, K.; Yoshida, H.; Zhang, H.; Sakka, Y. *J. Ceram. Soc. Jap.* 2012, *120,* 116-118.

Kim, B.-N.; Prajatelistia, E.; Han, Y.-H.; Son, H.-W.; Sakka, Y.; Kim, S. *Scripta Mater.* 2013, *69,* 366-369.

Kim, B. N.; Hiraga, K.; Jeong, A.; Hu, C.; Suzuki, T. S.; Yun, J.-D.; Sakka, Y. *J. Ceram. Soc. Jap.* 2014, *122,* 784-787.

Kim, B. N.; Dash, A.; Kim, Y. W.; Morita, K.; Yoshida, H.; Li, J.-G.; Sakka, Y. *J. Ceram. Soc. Jap.* 2016, *124,* 1141-1145.

Kingery, W. D.; Berg, M. *J. Appl. Phys.* 1955, *26,* 1205-1212.

Kingery, W. D.; Bowen, H. K.; Uhlmann, D. R. *Introduction to Ceramics;* Second edition; John Wiley & Sons: New York, NY, 1976; pp 448-532.

Klimke, J.; Trunec, M.; Krell, A. *J. Am. Ceram. Soc.* 2011, *94,* 1850-1858.

Kocjan, A.; Češnovar, M.; Vengust, D.; Kosmač, T.; Dakskobler, A.; Wei, W. C. *J. Am. Ceram. Soc.* 2016, *99,* 3090-3096.

Kodera, Y.; Hardin, C. L.; Garay, J. E. *Scripta Mater.* 2013, *69,* 149-154.

Korkmaz, E.; Sahin, F. *Acta Phys. Polon. A* 2017, *131,* 460-462.

Krell, A.; Hutzler, T.; Klimke, J. *J. Eur. Ceram. Soc.* 2009, *29,* 207-221.

Krell, A.; Klimke, J.; Hutzler, T. *Opt. Mater.* 2009, *31,* 1144-1150.

Kurosawa, S.; Pejchal, J.; Wakahara, S.; Yokota, Y.; Yoshikawa, A. *Radiat. Meas.* 2013, *56,* 155-158.

Kurosawa, S.; An, L.; Yamaji, A.; Suzuki, A.; Yokota, Y.; Shirasaki, K.; Yamamura, T.; Ito, A.; Goto, T.; Boulon, G.; Yoshikawa, A. *IEEE Trans. Nucl. Sci.* 2014, *61,* 316-319.

Kuwano, Y.; Suda, K.; Ishizawa, N.; Yamada, T. *J. Cryst. Crowth* 2004, *260,* 159-165.

Lallemant, L.; Fantozzi, G.; Garnier, V.; Bonnefont, G. *J. Eur. Ceram. Soc.* 2012, *32,* 2909-2915.

Lei, L.; Fu, Z.; Wang, H.; Lee, S. W.; Niihara, K. *Ceram. Int.* 2012, *38,* 23-28.

Li, W.; Mei, B.; Song, J.; Zhu, W.; Yi, G. *J. Alloys Comp.* 2016, *660,* 370-374.

Li, Z.; Thompson, B. C.; Dong, Z.; Khor, K. A. *Mater. Sci. Eng. C Mater. Biol. Appl.* 2016, *69,* 956-966.

Lin, D.; Fan, L.; Shi, Y.; Xie, J.; Lei, F.; Ren, D. *Opt. Mater.* 2017, *71,* 151-156.

Liu, C. J.; Sieckmann, E. F. *J. Appl. Phys.* 1966, *37,* 2450-2452.

Liu, J.; Shen, Z.; Yao, W.; Zhao, Y.; Mukherjee, A. K. *Nanotechnology* 2010, *21,* 75706 (5 pp).

Liu, Q.; Yuan, Y.; Li, J.; Liu, J.; Hu, C.; Chen, M.; Lin, L.; Kou, H.; Shi, Y.; Liu, W.; Chen, H.; Pan, Y.; Guo, J. *Ceram. Int.* 2014, *40,* 8539-8545.

Lucca, A.; Debourg, G.; Jacquemet, M.; Druon, F.; Balembois, F.; Georges, P.; Camy, P.; Doualan, J. L.; Moncongé, R. *Opt. Lett.* 2004, *29,* 2767-2769.

Malitson, I. H. *Appl. Opt.* 1963, *2,* 1103-1107.

Malitson, I. H.; Murphy Jr., F. V.; Rodney, W. S. *J. Opt. Soc. Am.* 1958, *48,* 72-73.

Meir, S.; Kalabukhov, S.; Froumin, N.; Dariel, M. P.; Frage, N. *J. Am. Ceram. Soc.* 2009, *92,* 358-364.

Morita, K.; Kim, B. N.; Hiraga, K.; Yoshida, H. *Scr. Mater.* 2008, *58,* 1114-1117.

Morita, K.; Kim, B.-N.; Yoshida, H.; Hiraga, K. *J. Am. Ceram. Soc.* 2009, *92,* 1208-1216.

Morita, K.; Kim, B.-N.; Yoshida, H.; Zhang, H.; Hiraga, K.; Sakka, Y. *J. Eur. Ceram. Soc.* 2012, *32,* 2303-2309.

Morita, K.; Kim, B.-N.; Yoshida, H.; Hiraga, K.; Sakka, Y. *Acta Mater.* 2015, *84,* 9-19.

Morita, K.; Kim, B.-N.; Yoshida, H.; Hiraga, K.; Sakka, Y.; Reimanis, I. *J. Am. Ceram. Soc.* 2015, *98,* 378-385.

Morita, K.; Kim, B.-N.; Yoshida, H.; Hiraga, K.; Sakka, Y. *J. Ceram. Soc. Jap.* 2015, *123,* 983-988.

Morita, K.; Kim, B.-N.; Yoshida, H.; Hiraga, K.; Sakka, Y. *J. Eur. Ceram. Soc.* 2016, *36,* 2961-2968.

Müller, M. M.; Kleebe, H. *J. Am. Ceram. Soc.* 2012, *95,* 3022-3024.

Munir, Z. A.; Anselmi-Tamburini, U.; Ohyanagi, M. *J. Mater. Sci.* 2006, *41,* 763-777.

Nakamura, F.; Kato, T.; Okada, G.; Kawaguchi, N.; Fukuda, K.; Yanagida, T. *Ceram. Int.* 2017, *43,* 604-609.

Nakamura, F.; Kato, T.; Okada, G.; Kawaguchi, N.; Fukuda, K.; Yanagida, T. *Opt. Mater.* 2017, *72,* 470-475.

Nakamura, F.; Kato, T.; Okada, G.; Kawaguchi, N.; Fukuda, K.; Yanagida, T. *J. Eur. Ceram. Soc.* 2017, *37,* 4919-4924.

Nakamura, F.; Kato, T.; Okada, G.; Kawaguchi, N.; Fukuda, K.; Yanagida, T. *J. Alloys Comp.* 2017, *726,* 67-73.

Nakamura, F.; Kato, T.; Okada, G.; Kawano, N.; Kawaguchi, N.; Fukuda, K.; Yanagida, T. *Mater. Res. Bull.* 2018, *98,* 83-88.

Nakayama, S.; Sakamoto, M. *J. Eur. Ceram. Soc.* 1998, *18,* 1413-1418.

Nam, S.; Lee, M.; Kim, B.-N.; Lee, Y.; Kang, S. *Ceram. Int.* 2017, *43,* 15352-15359.

Nigara, Y. *Jpn. J. Appl. Phys.* 1968, *7,* 404-408.

Olevsky, E.; Froyen, L. *J. Am. Ceram. Soc.* 2009, *92,* S122-S132.

Orrù, R.; Licheri, R.; Locci, A. M.; Cincotti, A.; Cao, G. *Mater. Sci. Eng. R* 2009, *63,* 127-287.

Pabst, W.; Hostaša, J. *Ceram. Silik.* 2013, *57,* 151-161.

Pabst, W.; Hostaša, J.; Esposito, L. *J. Eur. Ceram. Soc.* 2014, *34,* 2745-2756.

Pastrňák, J.; Roskovcová, L. *Phys. Status Solidi* 1966, *17,* K5-K8.

Penilla, E. H.; Kodera, Y.; Garay, J. E. *Mater. Sci. Eng. B* 2012, *177,* 1178-1187.

Petermann, K.; Fornasiero, L.; Mix, E.; Peters, V. *Opt. Mater.* 2002, *19,* 67-71.

Polyanskiy, M. N. *Refractive Index Database* (https://refractiveindex.info, accessed 1 March 2018).

Prahl, S. *Mie Scattering Calculator* (http://omlc.org./calc/mie_calc.html, accessed 1 March 2018).

Prakasam, M.; Viraphong, O.; Michau, D.; Veber, P.; Velázquez, M.; Shimamura, K.; Largeteau, A. *Ceram. Int.* 2013, *39,* 1307-1313.

Prakasam, M.; Viraphong, O.; Michau, D.; Largeteau, A. *Ceram. Int.* 2014, *40,* 1859-1864.

Prakasam, M.; Michau, D.; Viraphong, O.; Largeteau, A. *Adv. Appl. Ceram.* 2016, *115,* 333-341.

Prakasam, M.; Weill, F.; Lebraud, E.; Viraphong, O.; Buffiére, S.; Largeteau, A. *Int. J. Appl. Ceram. Technol.* 2016, *13,* 904-911.

Quach, D. V.; Groza, J. R.; Zavaliangos, A.; Anselmi-Tamburini, U. Fundamentals and applications of field/current assisted sintering, in

Sintering of Advanced Materials; Fang, Z. Z. (Ed.); Woodhead Publishing: Cambridge, UK, 2010; pp 249-274.

Querry, M. R. *Optical Constants;* Contractor report CRDC-CR-85034 (1985).

Rahaman, M. N. *Ceramic Prcoesses and Sintering;* Second edition; Marcel Dekker: New York, NY, 2003; pp 425-845.

Rahaman, M. N. Kinetics and mechanisms of densification, in *Sintering of Advanced Materials;* Fang, Z. Z. (Ed.); Woodhead Publishing: Cambridge, UK, 2010; pp 33-64.

Reimanis, I. E.; Kleebe, H. *Int. J. Mat. Res.* 2007, *98,* 1273-1278.

Ren, L.; Fu, Z.; Wang, Y.; Zhang, F.; Zhang, J.; Wang, W.; Wang, H. *Mater. Design* 2015, *83,* 753-759.

Rothman, A.; Kalabukhov, S.; Sverdlov, N.; Dariel, M. P.; Frage, N. *Int. J. Appl. Ceram. Technol.* 2014, *11,* 146-153.

Roussel, N.; Lallemant, L.; Durand, B.; Guillemet, S.; Ching, J.-Y. C.; Fantozzi, G.; Garnier, V.; Bonnefont, G. *Ceram. Int.* 2011, *37,* 3565-3573.

Rozenburg, K.; Reimanis, I. E.; Kleebe, H.; Cook, R. L. *J. Am. Ceram. Soc.* 2007, *90,* 2038-2042.

Rozenburg, K.; Reimanis, I. E.; Kleebe, H.; Cook, R. L. *J. Am. Ceram. Soc.* 2008, *36,* 444-450.

Salmang, H.; Scholze, H. *Keramik [Ceramics] – Volume I;* Sixth edition; Springer: Berlin, Germany; 1982; pp 117-177.

Schneider, H.; Schreuer, J.; Hildmann, B. *J. Eur. Ceram. Soc.* 2008, *28,* 329-344.

Schneider, H.; Komarneni, S. *Mullite;* Wiley-VCH: Weinheim, Germany, 2005, pp 1-487.

Schneider, H.; Fischer, R. X.; Schreuer, J.; Green, D. J. *J. Am. Ceram. Soc.* 2015, *98,* 2948-2967.

Shan, Y.; Wei, X.; Sun, X.; Xu, J.; Qin, Q.; Olevsky, E. A. *J. Mater. Res.* 2017, *32,* 3279-3285.

Shen, Z.; Johnsson, M.; Zhao, Z.; Nygren, M. *J. Am. Ceram. Soc.* 2002, *85,* 1921-1927.

Shen, Y.; Xu, J.; Tok, A.; Tang, D.; Khor, K. A.; Dong, Z. *J. Am. Ceram. Soc.* 2010, *93,* 3060-3063.

Shlyakhtina, A. V.; Ukshe, A. E.; Shcherbakova, L. G. *Elektrokhimiya [Russ. J. Electrochem.]* 2005, *41,* 265-269.

Shoja Razavi, R.; Ahsanzadeh-Vadeqani, M.; Barekat, M.; Naderi, M.; Hashemi, S. H.; Mishra, A. K. *Ceram. Int.* 2016, *42,* 7819-7823.

Siqueiros, J. M.; Machorro, R.; Regalado, L. E. *Appl. Optics* 1988, *27,* 2549-2553.

Sokol, M.; Kalabukhov, S.; Dariel, M. P.; Frage, N. *J. Eur. Ceram. Soc.* 2014, *34,* 4305-4310.

Sokol, M.; Kalabukhov, S.; Kasiyan, V.; Rothman, A.; Dariel, M. P.; Frage, N. *Opt. Mater.* 2014, *38,* 204-210.

Sokol, M.; Halabi, M.; Kalabukhov, S.; Frage, N. *J. Eur. Ceram. Soc.* 2017, *37,* 755-762.

Song, X.; Liu, X.; Zhang, J. *J. Am. Ceram. Soc.* 2006, *89,* 494-500.

Spina, G.; Bonnefont, G.; Palmero, P.; Fantozzi, G.; Chevalier, J.; Montanaro, L. *J. Eur. Ceram. Soc.* 2012, *32,* 2957-2964.

Stephens, R. E.; Malitson, I. H. *J. Res. Natl. Bur. Stand.* 1952, *49,* 249-252.

Stuer, M.; Zhao, Z.; Aschauer, U.; Bowen, P. *J. Eur. Ceram. Soc.* 2010, *30,* 1335-1343.

Swab, J. J.; Pavlacka, R.; Gilde, G.; Kilczewski, S.; Wright, J.; Harris, D.; Lara-Curzio, E. *J. Am. Ceram. Soc.* 2014, *97,* 592-600.

Su, H.; Johnson, D. L. *J. Am. Ceram. Soc.* 1996, *79,* 3211-3217.

Suárez, M.; Fernández, A.; Menéndez, J. L.; Torrecillas, R. *J. Nanomater.* 2009, *2009,* 1-5.

Suárez, M.; Fernández, A.; Menéndez, J. L.; Nygren, M.; Torrecillas, R.; Zhao, Z. *Ceram. Int.* 2013, *39,* 6669-6672.

Sudarsanan, K.; Mackie, P. E.; Young, R. A. *Mater. Res. Bull.* 1972, *7,* 1331-1338.

Taylor, G. F. GB Patent No. 385,629 (1932).

Taylor, G. F. US Patent No. 1,896,853 (1933).

Taylor, G. F. US Patent No. 1,896,854 (1933).

Tokita, M. *J. Soc. Powder Technol. Jap.* 1993, *30,* 790-840.

Tõldsepp, E.; Schoenstein, F.; Amamra, M.; Saar, R.; Feldbach, E.; Kanaev, A.; Kirm, M. *Ceram. Int.* 2016, *42,* 11709-11715.

Tran, T. B.; Hayun, S.; Navrotsky, A.; Castro, R. H. R.; Chaim, R. *J. Am. Ceram. Soc.* **2012**, *95,* 1185-1188.

Tropf, W. J.; Thomas, M. E. Magnesium Aluminum Spinel ($MgAl_2O_4$), in Palik, E. D. (Ed.) *Handbook of Optical Constants of Solids;* Academic Press: Orlando, FL, 1991; pp 883-897.

Trunec, M.; Klimke, J.; Shen, Z. J. *J. Eur. Ceram. Soc.* 2016, *36,* 4333-4337.

Van de Hulst, H. C. *Light Scattering by Small Particles;* Wiley: New York, NY, 1957 (reprint Dover Publications: New York, NY, 1981), pp 63-227.

Van der Laag, N. J.; Snel, M. D.; Magusin, P. C. M. M.; de With, G. *J. Eur. Ceram. Soc.* 2004, *24,* 2417-2424.

Wang, C.; Zhao, Z. *Scripta Mater.* 2009, *61,* 193-196.

Wang, C.; Zhao, Z. *Mater. Res. Bull.* 2010, *45,* 1127-1131.

Wang, S. F.; Zhang, J.; Luo, D. W.; Gu, F.; Tang, D. Y.; Dong, Z. L.; Tan, G. E. B.; Que, W. X.; Zhang, T. S.; Li, S.; Kong, L. B. *Prog. Solid State Chem.* 2013, *41,* 20-54.

Wang, D.; Shen, L.; Ran, S.; Huang, Q. *Scripta Mater.* **2014**, *92,* 31-34.

Wang, R.; Wang, Y.; Fu, Z.; Wang, H.; Wang, W.; Zhang, J.; Zhu, J. *J. Eur. Ceram. Soc.* 2016, *36,* 2153-2156.

Wang, R.; Liu, J.; Ji, W.; Wang, Y.; Fu, Z.; Wang, H.; Wang, W.; Zhang, J.; Zhu, J. *J. Alloys Comp.* 2017, *701,* 279-287.

Wang, P.; Yang, M.; Zhang, S.; Tu, R.; Goto, T.; Zhang, L. *J. Eur. Ceram. Soc.* 2017, *37,* 4103-4107.

Watanabe, Y.; Ikoma, T.; Monkawa, A.; Suetsugu, Y.; Yamada, H.; Tanaka, J.; Moriyoshi, Y. *J. Am. Ceram. Soc.* 2005, *88,* 243-245.

Wemple, S. H.; Didomenico Jr., M.; Camlibel, I. *J. Phys. Chem. Solids* 1968, *29,* 1797-1803.

Wood, D. L.; Nassau, K.; Kometani, T. Y. *Appl. Opt.* 1990, *29,* 2485-2488.

Wu, Y. J.; Kimura, R.; Uekawa, N.; Kakegawa, K.; Sasaki, Y. *Jpn. J. Appl. Phys.* 2002, *41,* L219-L221.

Wu, Y. J.; Li, J.; Kimura, R.; Uekawa, N.; Kakegawa, K. *J. Am. Ceram. Soc.* 2005, *88,* 3327-3331.

Wu, Y. J.; Wang, N.; Wu, S. Y.; Chen, X. M. *J. Am. Ceram. Soc.* 2011, *94,* 1343-1345.

Xing, B.; Cheng, Z.; Wang, C.; Zhao, Z. *Opt. Mater.* 2017, *71,* 13-16.

Xiong, Y.; Fu, Z. Y.; Wang, H.; Wang, Y. C.; Zhang, Q. J. *Mater. Sci. Eng. B* **2005**, *123,* 57-62.

Xiong, Y.; Fu, Z.; Wang, H.; Wang, W.; Zhang, J.; Zhang, Q.; Lee, S. W.; Niihara, K. *J. Am. Ceram. Soc.* 2007, *90,* 1647-1649.

Xiong, Y.; Hu, J.; Shen, Z.; Pouchly, V.; Maca, K.; Bordia, R. *J. Am. Ceram. Soc.* 2011, *94,* 4269-4273.

Xu, S.; Li, J.; Kou, H.; Shi, Y.; Pan, Y.; Guo, J. *Ceram. Int.* 2015, *41,* 3312-3317.

Yanagida, T. *Opt. Mater.* 2013, *35,* 1987-1992.

Yanagida, T.; Yoshikawa, A.; Ikesue, A.; Kamada, K.; Yokota, Y. *IEEE Trans. Nucl. Sci.* 2009, *56,* 2955-2959.

Yanagisawa, O.; Kuramoto, H.; Matsugi, K.; Komtasu, M. *Mat. Sci. Eng. A* 2003, *350,* 184-189.

Yang, W.; Hojo, J.; Enomoto, N.; Tanaka, Y.; Inada, M. *Mater. Lett.* 2013, *96,* 155-157.

Yong, X.; Ping, F.; Baohua, Z.; Juan, G.; Lin, Z.; Xuehua, W. *Mater. Lett.* 2014, *123,* 142-144.

Yoshida, H.; Morita, K.; Kim, B.-N.; Hiraga, K.; Yamanaka, K.; Soga, K.; Yamamoto, T. *J. Am. Ceram. Soc.* 2011, *94,* 3301-3307.

Yun, J.; Son, H.; Prajatelistia, E.; Han, Y.-H.; Kim, S.; Kim, B.-N. *Adv. Appl. Ceram.* 2014, *113,* 67-72.

Zelmon, D. E.; Small, D. L.; Page, R. *Appl. Opt.* 1998, *37,* 4933-4935.

Zerarga, F.; Bouhemadou, A.; Khenata, R.; Bin-Omran, S. *Solid State Sci.* 2011, *13,* 1638-1648.

Zhang, G.; Wang, Y.; Fu, Z.; Wang, H.; Wang, W.; Zhang, J.; Lee, S. W.; Niihara, K. *J. Eur. Ceram. Soc.* 2009, *29,* 2705-2711.

Zhang, H. B.; Kim, B. N.; Morita, K.; Yoshida, H.; Lim, J. H.; Hiraga, K. *J. Alloys Comp.* 2010, *508,* 196-199.

Zhang, H.; Kim, B. N.; Morita, K.; Yoshida, H.; Lim, J.-H.; Hiraga, K. *J. Am. Ceram. Soc.* 2011, *94,* 2981-2986.

Zhang, H.; Li, Z.; Kim, B. N.; Morita, K.; Yoshida, H.; Hiraga, K.; Sakka, Y.; Blendell, J. *J. Am. Ceram. Soc.* 2011, *94,* 2739-2741.

Zhang, H.; Kim, B. N.; Morita, K.; Yoshida, H.; Hiraga, K.; Sakka, Y.; Ballato, J. *J. Am. Ceram. Soc.* 2011, *94,* 3206-3210.

Zhang, H.; Kim, B. N.; Morita, K.; Yoshida, H.; Hiraga, K.; Sakka, Y. *Sci. Technol. Adv. Mater.* 2011, *12,* 055003 (6 pp).

Zhang, H.; Li, Z.; Kim, B.-N.; Morita, K.; Yoshida, H.; Hiraga, K.; Sakka, Y. *J. Nanomater.* 2012, *2012,* 1-5.

Zhang, X.; Fan, G.; Lu, W.; Chen, Y.; Ruan, X. *J. Eur. Ceram. Soc.* 2016, *36,* 2767-2772.

Zhou, Y.; Hirao, K.; Yamauchi, Y.; Kanzaki, S. *J. Eur. Ceram. Soc.* 2004, *24,* 3465-3470.

APPENDIX A. REFRACTIVE INDICES AND THEORETICAL MAXIMUM TRANSMITTANCES

Table A1 lists effective refractive indices (in the case of non-cubic, i.e., optically anisotropic, crystallites calculated via Equation 7) and theoretical maximum transmittances calculated from these values via Equation 4 for wavelengths 600 nm and 1064 nm, i.e., in the middle of the visible part of the electromagnetic spectrum and at the infrared wavelength generated by a Nd:YAG laser. The refractive indices have been taken from the literature sources cited in the last column of this table. Since for this calculation the maximum transmittances have been calculated from effective refractive indices (i.e., average values for polycrystalline materials with randomly oriented crystallites), the effect of scattering at the grain boundaries of non-cubic crystallites is not taken into account in this calculation. That means, for polycrystalline materials with non-cubic crystallites this maximum transmittance can never be attained, even when second-phase inclusions are completely absent.

Table A1. Effective refractive indices and theoretical maximum transmittances of materials that have been successfully prepared via ECAS and are treated in this chapter

Name	Refractive index at 600 nm	Refractive index at 1064 nm	Maximum transmittance at 600 nm	Maximum transmittance at 1064 nm	Ref.
Alumina	1.7635	1.7506	85.8	86.1	1
Aluminium nitride	2.1758	2.1506	75.9	76.5	2
Aluminium oxynitride	1.7700	1.7480	85.7	86.2	3
Barium titanate	2.3974	-	71.1	-	4
Calcium fluoride	1.4336	1.4285	93.9	94.0	5
Calcium oxide	1.8330	-	84.1	-	6
Hydroxyapatite	1.6475	1.6475	88.7	88.7	7
Lutetium aluminum garnet	1.8318	1.8249	84.1	84.3	8
Lutetium oxide	1.9200	1.9106	81.9	82.2	9
Magnesium aluminate	1.7155	1.7019	87.0	87.4	10
Magnesium fluoride	1.3775	1.3732	95.1	95.2	11
Magnesium oxide	1.7367	1.7216	86.5	86.9	12
Mullite	1.6570	1.6570	88.5	88.5	13
Sphalerite	2.3625	2.2883	71.8	73.4	14
Strontium fluorapatite	1.6310	1.6310	89.1	89.1	15
Strontium titanate	2.4046	2.3103	70.9	72.9	16
Yttria	1.9289	1.9011	81.7	82.4	17
Yttrium aluminum garnet	1.8318	1.8147	84.1	84.5	18
Zinc aluminate	1.6600	1.6369	88.4	89.0	19
Zinc oxide	2.0010	1.9475	80.0	81.3	20
Zirconia (FSZ)	2.2709	2.2320	73.8	74.6	21

References: 1. Malitson et al. 1958, 2. Pastrňák & Roskovcová 1966, 3. Hartnett et al. 1998, 4. Wemple et al. 1968, 5. Malitson 1963, 6. Liu & Sieckmann 1966, 7. Bernard & Rost 1992, 8. Kuwano et al. 2004, 9. Kaminskii et al. 2008, 10. Tropf & Thomas 1997, 11. Dodge 1984, 12. Stephens & Malitson 1952, 13. Schneider et al. 2015, 14. Debenham 1984, 15. Sudarsanan et al. 1972, 16. Liu et al. 2010, 17. Nigara 1968, 18. Zelmon et al. 1998, 19. Zeraga et al. 2011, 20. Bond 1965, 21. Wood et al. 1990.

APPENDIX B. DETAILS OF ECAS REGIMES AND OPTICAL PROPERTIES

Based on the literature review given in this chapter, this appendix section lists the details on composition, ECAS processing regimes used and optical properties obtained for the individual materials cited by different authors. Complementary information can be found in the main text.

Table B1. ECAS schedules used and optical properties achieved for specimens of different simple oxide polycrystalline materials according to the literature (superscript a = in-line transmittance, superscript b = real in-line transmittance, superscript c = total forward transmittance)

Name	Sintering temperature [°C]	Heating rate [°C/min]	Highest pressure [MPa]	Dwell time [min]	Post-annealing [°C/h]	Doping and sintering additives	Optical properties (best result)	Thickness [mm]	Ref.
Alumina	1150	8–25	80	0	-	-	47% at 600 nm[a]	0.88	1
Alumina	1300–1680	170–1000	90	5	1250/24	-	86% at 4000 nm[a]	1.00	2
Alumina	1100–1550	2	80	20	-	-	33% at 640 nm[a]	1.00	3
Alumina	1150	2–100	80	20	-	-	46% at 640 nm[a]	0.90	4
Alumina	1300	8	80	60	-	-	29% at 640 nm[a]	0.80	5
Alumina	1100–1350	8–600	80	0–60	-	-	38% at 640 nm[b]	0.80	6
Alumina	900–1000	70	500	10	-	-	64% at 645 nm[b]	0.80	7
Alumina	1200–1400	100	80	3	-	-	53% at 640 nm[a]	0.50	8
Alumina	1100–1500	2	80	20	-	0.03–0.05 wt% MgO	38% at 640 nm[b]	1.00	9
Alumina	1350–1400	94–100	50–100	n/a	-	450 ppm $Mg^{2+}/Y^{3+}/La^{3+}$	57% at 640 nm[a]	0.80	10

Name	Sintering temperature [°C]	Heating rate [°C/min]	Highest pressure [MPa]	Dwell time [min]	Post-annealing [°C/h]	Doping and sintering additives	Optical properties (best result)	Thickness [mm]	Ref.
Alumina	1350	50–200	100	5	-	0.05–1 wt% Cr_2O_3	85% at 2000 nm[b]	0.80	11
Alumina	1250–1430	50	80	2–20	-	0.5 wt% CeO_2	16% at 640 nm[b] 71% at 4000 nm[b]	n/a	12
Alumina	1150	70–100	80	30–60	-	-	19% at 645 nm[a]	0.80	13
Alumina	1180–1280	10	80	10	-	100–500 ppm $La^{3+}/Mg^{2+}/Zr^{4+}$	48% at 640 nm[b]	0.88	14
Alumina	1250–1430	50	80	2–20	-	0.3–1 wt% CeO_2	42% at 640 nm[b]	0.80	15
Alumina	1000–1300	10	50–400	20	-	0.03 wt% MgO	64% at 640 nm[a]	1.00	16
Alumina	1130–1230	1–10	80	0	-	-	53% at 640nm[b]	0.88	17
Alumina	1400	100	100	3	-	-	52% at 640 nm[a]	0.80	18
Alumina	1350	100	72	5	-	0–0.5 wt% AlF_3	45% at 650 nm[c]	1.00	19
Alumina	1000	5–50	80–160	7–20	-	-	39% at 640 nm[a]	1.00	20
Alumina	1150–1400	5–100	40–100	0–120	-	-	51% at 640 nm[a]	1.00	21
Alumina	1550	8–50	80	20	-	-	10% at 640 nm[a]	0.80	22
Alumina	900–1400	100	70–1000	20–30	-	-	75% at 640 nm[a]	0.50	23
Calcia	1600	70	80	60	-	-	-	n/a	24
Lutecia	1000–1500	10	20–100	45–600	-	-	40% at 640 nm[a] 62% at 2000 nm[a]	1.00	25
Lutecia	1450	2–200	100	45	-	-	60% at 550 nm[a]	1.00	26
Lutecia	1450	10 – 200	100	45	1150/12	-	72% at 550 nm[a] 82% at 2000 nm[a]	1.00	27

Table B1. (Continued)

Name	Sintering temperature [°C]	Heating rate [°C/min]	Highest pressure [MPa]	Dwell time [min]	Post-annealing [°C/h]	Doping and sintering additives	Optical properties (best result)	Thickness [mm]	Ref.
Lutecia	1400–1800	5–200	40–100	5	1200/12	10% Yb^{3+}	52% at 640 nm[a] 64% at 2000 nm[a]	2.00	28 29
Lutecia	1400	50	130	15	-	0.5 at% Nd^{3+}	81% at 1060 nm[a]	2.00	30
Lutecia	1450	10–200	100	5	850/6	1 at% Nd^{3+} 0–0.5 wt% LiF	82% at 1080 nm[a]	1.00	31 32
Lutecia	1450	10–200	100	5	-	1 at% Nd^{3+} 1 wt% LiF	80% at 640 nm[a]	1.00	33
Lutecia	1450	10–200	100	5	850–1050/6	0–5 mol% Nd^{3+} 0.2 wt% LiF	80% at 1080 nm[a]	1.05	34
Lutecia	1450	10–138	50	40	1400/2	5 at% Yb^{3+}	76% at 1080 nm[a]	1.00	35
Lutecia	1450	2–200	100	45	850/6	1–5 at% Nd^{3+}/Eu^{3+} 0.05–0.5 wt% LiF	78% at 800 nm[a] 82% at 1080 nm[a]	1.00	36
Magnesia	700–825	n/a	150	5	-	-	37% at 640 nm[a]	1.05	37
Magnesia	650–800	25–100	300–500	10–30	-	0.4 mol% CaO	22% at 640 nm[a] 71% at 2000 nm[a]	0.80	38
Magnesia	1400	17	80	60	-	0.001–0.1 at% Cr^{3+}	27% at 640 nm[a]	n/a	39
Magnesia	1400	17	80	60	-	0.001–0.1% C	22% at 640 nm[a]	n/a	40
Magnesia	1500	n/a	80	60	-	0.001–1 at% Mn^{2+}	33% at 640 nm[a]	1.50	41
Magnesia	1500	17	80	60	-	0.001–0.1 mol% Ce^{4+}	40% at 640 nm[a]	0.32	42

Name	Sintering temperature [°C]	Heating rate [°C/min]	Highest pressure [MPa]	Dwell time [min]	Post-annealing [°C/h]	Doping and sintering additives	Optical properties (best result)	Thickness [mm]	Ref.
Magnesia	800–1100	10–300	30	5	-	0–2 wt% LiF	17% at 640 nm[a] 77% at 2000 nm[a]	1.50	43
Magnesia	1500	17	80	60	-	0–0.5 at% Tb^{3+}	16% at 640 nm[a]	1.10	44
Yttria	850–1050	2–50	80	60	-	-	6% at 400 nm[a] 46% at 800 nm[a]	1.00	45
Yttria	1050	20	200–300	60	900/3	-	68% at 700 nm[a]	1.00	46
Yttria	1100–1550	10–200	100	45	850–1150/6	-	55% at 550 nm[a] 82% at 2000 nm[a]	1.00	47
Yttria	1050–1150	112.5	60	25–30	800–1000/3–6	-	6% at 640 nm[a] 51% at 2000 nm[a]	1.00	48
Yttria	1200–1600	100	100	20	1200/10	-	83% at 2500 nm[a]	2.50	49
Yttria	1500	100	100	8	1300/10	-	80% at 2500 nm[a]	3.00	50
Yttria	1200–1600	100–280	100	20	1200/10	-	11% at 400 nm[a] 54% at 800 nm[a]	2.50	51
Yttria	1500	100–280	100	20	1200/10	3 at% Zr^{4+}	24% at 400 nm[a] 83% at 5000 nm[a]	2.50	52
Yttria	1300	50	80–90	45	1050/6	-	60% at 5000 nm[a]	3.50	53
Yttria	1250–1350	10	45	5–30	1200/n/a	0–20 at% Sc^{3+}	82% at 5300 nm[a]	1.00	54
Yttria	1400–1550	10–100	6	30	950/8	3 at% Sm^{3+} 0–0.5 wt% LiF	75% at 609 nm[a]	1.70	55
Yttria	1300	10–100	100	40	1050/6	-	77% at 2000 nm[a]	0.50	56

Table B1. (Continued)

Name	Sintering temperature [°C]	Heating rate [°C/min]	Highest pressure [MPa]	Dwell time [min]	Post-annealing [°C/h]	Doping and sintering additives	Optical properties (best result)	Thickness [mm]	Ref.
Yttria	1300–1350	50	90	15	1050–1250/6	-	80% at 5000 nm[a]	3.50	57
Yttria	1050–1250	5	82.7	60	1100/1	0–50 at% Yb^{3+}	78% at 1030 nm[a]	0.80	58
Zinc oxide	900–1100	5–200	40–100	20	700/5	-	69% at 640 nm[a]	2.00	59
Zinc oxide	850	10–100	80	10	-	-	12% at 950 nm[a]	1.00	60
Zirconia (FSZ)	1200	n/a	106	10–12	750/2–24	-	56% at 630 nm[a]	n/a	61
FSZ	1050–1200	5–50	400	2–180	900/4	-	48% at 640 nm[a]	1.00	62
FSZ	1000–1200	10	400	10	900/4	-	43% at 640 nm[a]	1.30	63
FSZ	1200–1350	60–100	50	5	-	-	28% at 640 nm[a]	0.60	64
FSZ	1150–1400	2–200	40–100	20	-	-	41% at 640 nm[a]	1.00	65
Zirconia (TZP)	1050	50	400	10	900/2	-	63% at 3000 nm[a]	1.50	66
TZP	1100	100–200	100	60	-	-	-	n/a	67
TZP	1050	10	400	10	900/4	0–0.3 mol% Al_2O_3	7% at 640 nm[a]	1.00	68

References: 1. Kim et al. 2007, 2. Jiang et al. 2008, 3. Kim et al. 2009a, 4. Kim et al. 2009b, 5. Aman et al. 2009, 6. Aman et al. 2010, 7. Grasso et al. 2010, 8. Jin et al. 2010, 9. Kim et al. 2010b, 10. Stuer et al. 2010, 11. Wang & Zhao 2010, 12. Alvarez-Clemares et al. 2010, 13. Grasso et al. 2011, 14. Roussel et al. 2011, 15. Alvarez-Clemares et al. 2012, 16. Kim et al. 2012, 17. Lallemant et al. 2012, 18. Suárez et al. 2013, 19. Wang et al. 2014, 20. Kim et al. 2016, 21. Prakasam et al. 2016a, 22. Tõldsepp et al. 2016, 23. Ghanizadeh et al. 2017, 24. Kato et al. 2017a, 25. An et al. 2011b, 26. An et al. 2011c, 27. An et al. 2011d, 28. Prakasam et al. 2013, 29. Prakasam et al. 2014, 30. Boulesteix et al. 2014, 31. An et al. 2014a, 32. An et al. 2014b, 33. Kurosawa et al. 2014, 34. Alombert-Goget et al. 2015, 35. Kijko et al. 2015, 36. Ito et al. 2016, 37. Chaim et al. 2004, 38. Tran et al. 2012, 39. Kato et al. 2016a, 40. Kato et al. 2016b, 41. Kato et al. 2016c, 42. Kato et al. 2017b, 43. Jiang et al. 2017, 44. Kawano et al. 2017, 45. Yoshida et al. 2011, 46. Zhang et al. 2011c, 47. An et al. 2012a, 48. Attia et al. 2015, 49. Shoja Razavi et al. 2016, 50. Ghaderi et al. 2016, 51. Ahmadi et al. 2016, 52. Ahsanzadeh-Vadeqani & Shoja Razavi 2016, 53. Irom et al. 2016, 54. Du et al. 2017, 55. Hu et al. 2017, 56. Korkmaz & Sahin 2017, 57. Irom et al. 2018. 58. Furuse et al. 2018, 59. Prakasam et al. 2014, 60. Lin et al. 2017, 61. Alaniz et al. 2009, 62. Zhang et al. 2010, 63. Zhang et al. 2011d, 64. Lei et al. 2012, 65. Prakasam et al. 2016b, 66. Zhang et al. 2011b, 67. Xiong et al. 2011, 68. Zhang et al. 2012.

Table B2. ECAS schedules used and optical properties achieved for specimens of different mixed oxide polycrystalline materials according to the literature (superscript a = in-line transmittance, superscript b = real in-line transmittance, superscript c = total forward transmittance)

Name	Sintering temperature [°C]	Heating rate [°C/min]	Highest pressure [MPa]	Dwell time [min]	Post-annealing [°C/h]	Doping and sintering additives	Optical properties (best result)	Thickness [mm]	Ref.
Barium titanate	900	100	200	3	750/3.3	-	14% at 640 nm[a] 50% at 2400 nm[a]	0.84	1
Barium-stroncium titanate	1000	20–100	50	5	1000/2	-	74% at 633 nm[a]	0.30	2
Ba-Zr-Ce-Y perovskite	1350	100	23	5	1350–1500/10	-	71% at 2000 nm[a]	0.80	3
La-doped lead zirconate titanate	700–900	33–100	29	10	700–1000/1	-	31% at 700 nm[a]	0.50	4

Table B2. (Continued)

Name	Sintering temperature [°C]	Heating rate [°C/min]	Highest pressure [MPa]	Dwell time [min]	Post-annealing [°C/h]	Doping and sintering additives	Optical properties (best result)	Thickness [mm]	Ref.
Lutetium aluminum garnet	1200–1650	10–200	100	45	1150/12	-	78% at 2000 nm[a]	1.00	5
Lutetium hafnium oxide	1450	10–200	100	45	850/6	-	71% at 1100 nm[a] 78% at 2000 nm[a]	1.00	6
Lutetium niobate	1500	10–200	100	45	850/6	-	68% at 550 nm[a]	1.10	7
Lutetium niobate	1200–1550	10–200	100	45	750–1250/6	-	63% at 550 nm[a]	1.00	8
Lutetium titanate pyrochlore	1450	10–200	100	45	750/6	-	40% at 550 nm[a] 72% at 2200 nm[a]	0.80	9
Lutetium titanate pyrochlore	1000–1550	10–200	100	45	650–1550/60	-	57% at 550 nm[a] 74% at 2000 nm[a]	1.00	10
Lutetium titanate pyrochlore	1450	10–200	100	45	750/6	-	34% at 550 nm[a] 74% at 2000 nm[a]	1.00	11
Magnesium aluminate spinel (MAS)	1600	n/a	64	n/a	-	1 wt% LiF	75% at 1000 nm[a]	2.70	12
MAS	1300	2–100	80	20	-	-	47% at 550 nm[a]	1.80	13
MAS	n/a	100	n/a	n/a	-	-	47% at 550 nm[a]	1.19	14
MAS	900–1600	200	64	0–60	-	1 wt% LiF	78% at 800 nm[a]	2.10	15
MAS	1275–1500	2–100	80	20	-	-	50% at 550 nm[a] 70% at 2000 nm[a]	1.80	16
MAS	1300	50–200	100	3	-	-	51% at 550 nm[b]	1.79	17

Name	Sintering temperature [°C]	Heating rate [°C/min]	Highest pressure [MPa]	Dwell time [min]	Post-annealing [°C/h]	Doping and sintering additives	Optical properties (best result)	Thickness [mm]	Ref.
MAS	1325	10	80	20	-	-	61% at 600 nm[a]	1.80	18
MAS	1250–1320	1–100	72	0	-	-	74% at 550 nm[b]	1.80	19
MAS	1300	100	80	20	-	-	35% at 550 nm[a] 74% at 2000 nm[a]	1.80	20
MAS	1250–1375	10–300	80	20	-	-	70% at 550 nm[a]	1.00	21
MAS	1300–1500	2–100	63–95	10–120	-	0–1 wt% LiF	80% at 500 nm[a]	1.50	22
MAS	1150–1300	5–50	150–400	2–30	-	-	83% at 600 nm[a]	1.00	23
MAS	1300	10–300	80	20	800–1300/1	-	75% at 550 nm[a]	1.00	24
MAS	1325	10–300	80	10	-	0–3 at% Zn^{2+}	70% at 550 nm[a] 87% at 2000 nm[a]	1.00	25
MAS	1300	100	70	0	-	-	83% at 3000 nm[a]	3.00	26
MAS	1450	n/a	64	15	-	-	73% at 550 nm[a]	2.50	27
MAS	1300	10	80	20	1100–1300/1	-	60% at 550 nm[a]	1.80	28
MAS	1300–1400	1–100	73	0	-	-	70% at 550 nm[a] 78% at 1100 nm[a]	1.00	29
MAS	1300	20–100	130	15	-	0.1–0.2 wt% Nd^{3+}	48% at 550 nm[a]	1.70	30
MAS	1250	10	80	20	-	-	56% at 550 nm[b] 75% at 1500 nm[b]	2.00	31
MAS	1000	5	1000	n/a	-	-	80% at 550 nm[a]	1.50	32
PZ-PT-PZN	900	33–100	29	10	900–1200/1	-	13% at 633 nm[a]	0.30	33

Table B2. (Continued)

Name	Sintering temperature [°C]	Heating rate [°C/min]	Highest pressure [MPa]	Dwell time [min]	Post-annealing [°C/h]	Doping and sintering additives	Optical properties (best result)	Thickness [mm]	Ref.
Strontium hafnium oxide	1700	17–200	100	45	-	0 – 4 at% Ce^{4+}/Al^{3+}	10% at 640 nm[a]	1.00	34
Strontium titanate	950	100	200	3	750/3.3	-	4% at 640 nm[a] 94% at 2400 nm[a]	0.87	1
Yttrium aluminum garnet (YAG)	1250–1500	100–200	100	0	-	-	-	n/a	35
YAG	1200–1500	100–200	50–100	3–9	-	-	14% at 500 nm[a]	0.50	36
YAG	1500	100	50	3	-	-	60% at 680 nm[a]	0.80	37
YAG	1300	n/a	64	2	-	0–4 wt % LiF	76% at 640 nm[a]	3.00	38
YAG	1200–1400	2–100	64	1–120	-	1 at% Nd^{3+} 0.25 wt% LiF	80% at 1000 nm[a]	2.00	39 40
YAG	1400–1550	200	70	10	950/12	-	55% at 1064 nm[a]	0.90	41
YAG	1200–1400	8–100	74	15	900/1	-	66% at 600 nm[a]	1.00	42
YAG	700–1100	100	0.5–80	0–50	-	-	44% at 680 nm[a] 66% at 1000 nm[a]	1.00	43
YAG	1250–1390	10–100	80	20	-	0.25 wt% LiF	72% at 600 nm[a] 82% at 1064 nm[a]	1.00	44

Name	Sintering temperature [°C]	Heating rate [°C/min]	Highest pressure [MPa]	Dwell time [min]	Post-annealing [°C/h]	Doping and sintering additives	Optical properties (best result)	Thickness [mm]	Ref.
YAG	950	100	80	30	-	-	32% at 680 nm[a] 60% at 1000 nm[a]	1.00	45
Zinc aluminate spinel	1260–1275	10	n/a	15	-	0–0.5 wt% TEOS	64% at 550 nm[a] 85% at 2000 nm[a]	1.00	46
Zinc aluminate spinel	1200	10–50	80	20	1100/3	-	77% at 640 nm[a]	1.00	47
Zinc aluminate spinel	1240–1280	10–100	100	15	-	-	64% at 550 nm[a] 87% at 2500 nm[a]	1.00	48

References: 1. Liu et al. 2010, 2. Wu et al. 2011, 3. Xing et al. 2017, 4. Wu et al. 2005, 5. An et al. 2012c, 6. An et al. 2013b, 7. An et al. 2011f, 8. An et al. 2013a, 9. An et al. 2011a, 10. An et al. 2011e, 11. An et al. 2012b, 12. Frage et al. 2007, 13. Morita et al. 2008, 14. Bernard-Granger et al. 2009, 15. Meir et al. 2009, 16. Morita 2009, 17. Wang & Zhao 2009, 18. Kim et al. 2010a, 19. Bonnefont et al. 2012, 20. Morita et al. 2012, 21. Fu et al. 2013, 22. Rothmann et al. 2014, 23. Sokol et al. 2014a, 24. Fu et al. 2014, 25. Fu et al. 2015, 26. Biswas et al. 2016, 27. Alekseev et al. 2016, 28. Morita et al. 2016, 29. Benaissa et al. 2016, 30. Boulesteix et al. 2017, 31. Nam et al. 2017, 32. Sokol et al. 2017, 33. Wu et al. 2002, 34. Kurosawa et al. 2013, 35. Chaim et al. 2006, 36. Chaim et al. 2007, 37. Suárez et al. 2009, 38. Frage et al. 2010, 39. Frage et al. 2012, 40. Sokol et al. 2014b, 41. Penilla et al. 2012, 42. Spina et al. 2012, 43. Wang et al. 2016, 44. Zhang et al. 2016, 45. Wang et al. 2017, 46. Yong et al. 2014, 47. Kim et al. 2014, 48. Fu et al. 2017.

Table B3. ECAS schedules used and optical properties achieved for specimens of different polycrystalline phosphates and silicates according to the literature (superscript a = in-line transmittance, superscript b = real in-line transmittance, superscript c = total forward transmittance)

Name	Sintering temperature [°C]	Heating rate [°C/min]	Highest pressure [MPa]	Dwell time [min]	Post-annealing [°C/h]	Doping and sintering additives	Optical properties (best result)	Thickness [mm]	Ref.
Hydroxyapatite (HA)	800–1000	25	10	10	-	-	-	n/a	1
HA	1200	50	50	10	-	-	70% at 700 nm[a]	1.00	2
HA	700–1000	50–100	75–500	3–5	-	-	70% at 450 nm[a]	0.70	3
HA	750–1050	200	100	3–5.25	-	-	45% at 700 nm[a]	2.00	4
HA	900–1000	20	80	10	-	-	70% at 400 nm[a]	1.00	5
HA	900–1000	20	80	10	-	-	68% at 700 nm[a]	2.70	6
HA	900–1000	20	80	10			70% at 400 nm[a]	1.00	7
HA	900–1100	100–200	50	3	-	-	69% at 700 nm[a] 84% at 1300 nm[a]	2.00	8
Lanthanum oxyapatite	1200–1500	50–300	100	3	1000/12	-	-	0.50	9
Mullite	1450–1600	100	30	5–20	-	-	75% at 2500 nm[a] 82% at 4300 nm[a]	0.60	10
Mullite	1250–1450	100	80	10	-	-	83% at 2500 nm[a]	0.60	11
Mullite	1370	100	15	2–20	-	-	22% at 640 nm[a] 77% at 2000 nm[a]	1.00	12

Name	Sintering temperature [°C]	Heating rate [°C/min]	Highest pressure [MPa]	Dwell time [min]	Post-annealing [°C/h]	Doping and sintering additives	Optical properties (best result)	Thickness [mm]	Ref.
Strontium fluorapatite	1050	50	100	8	-	20 mol% Yb^{3+}	50% at 500 nm[a] 74% at 1000 nm[a]	2.00	13
Strontium yttrium oxyapatite	1400–1500	100–300	23	3	1400/4	-	3% at 800 nm[a]	0.80	14
Tricalcium phosphate	800–1000	25	60	0	-	-	-	n/a	15

References: 1. Ioku et al. 2002, 2. Watanabe et al. 2005, 3. Eriksson et al. 2011, 4. Chaudhry et al. 2011, 5. Kim et al. 2013, 6. Yun et al. 2014, 7. Han 2016, 8. Li et al. 2016b, 9. Chesnaud et al. 2007, 10. Zhang et al. 2009, 11. Ren et al. 2015, 12. Kocjan et al. 2016, 13. Chen et al. 2013, 14. Shen et al. 2010, 15. Kawagoe et al. 2004.

Table B4. ECAS schedules used and optical properties achieved for specimens of different polycrystalline non-oxides and composites according to the literature (superscript a = in-line transmittance, superscript b = real in-line transmittance, superscript c = total forward transmittance)

Name	Sintering temperature [°C]	Heating rate [°C/min]	Highest pressure [MPa]	Dwell time [min]	Post-annealing [°C/h]	Doping and sintering additives	Optical properties (best result)	Thickness [mm]	Ref.
Aluminium nitride	1800	100–550	25	4–20	-	-	-	n/a	1
Aluminium nitride	1800	100	30	15	-	0–3 wt% CaF_2	55% at 5500 nm[a]	0.40	2

Table B4. (Continued)

Name	Sintering temperature [°C]	Heating rate [°C/min]	Highest pressure [MPa]	Dwell time [min]	Post-annealing [°C/h]	Doping and sintering additives	Optical properties (best result)	Thickness [mm]	Ref.
Aluminium oxynitride	1500–1700	30	30	1–30	-	n/a % Dy^{3+}/Mg^{2+}	18% at 5680 nm[a]	n/a	3
Aluminium oxynitride	1350–1500	100	40	10	-	0.5 wt% Y_2O_3	77% at 3900 nm[a]	1.20	4
Calcium fluoride	600	100	30	5	-	5 at% Yb^{3+}	61% at 400 nm[a] 83% 1200 nm[a]	2.00	5
Calcium fluoride	1100	100	100	10	-	-	54% at 300 nm[a] 86% at 1100 nm[a]	1.00	6
Calcium fluoride	1070	80–90	70	15	-	0.01–0.1 mol% Eu^{2+}	34% at 640 nm[a]	1.00	7
Calcium fluoride	980	70–87	70	15	-	1–10 mol% Nd^{3+}	12% at 640 nm[a]	0.73	8
Magnesium fluoride	700	100	70	7	-	0.01–1 mol% Ce^{3+}	15% at 640 nm[a]	0.50	9
Magnesium fluoride	650	10–100	70	15	-	0.01–0.1 mol% Eu^{2+}	26% at 640 nm[a]	1.03	10

Name	Sintering temperature [°C]	Heating rate [°C/min]	Highest pressure [MPa]	Dwell time [min]	Post-annealing [°C/h]	Doping and sintering additives	Optical properties (best result)	Thickness [mm]	Ref.
Magnesium fluoride	750	100	70	7	-	0.01–1 mol% Gd^{3+}	24% at 640 nm[a]	1.00	11
Silicon alumina oxynitride	1850	100	30	5	-	3 wt% MgO	66% at 3470 nm[a]	0.50	12
Silicon alumina oxynitride	1850	100	30	5	-	3 wt% MgO	9% at 640 nm[a] 66% at 3500 nm[a]	0.20	13
Sphalerite	850	5–100	80	5	-	-	68% at 5000 nm[a]	1.50	14
Yttria-magnesium oxide composite	1200–1600	80–200	80	3–8	1000/15	-	85% at 4000 nm[a]	1.00	15
Yttria-magnesium oxide composite	1000–1300	100	50	5	1000/15	-	58% at 4000 nm[a]	1.22	16

References: 1. Fu et al. 2004, 2. Xiong et al. 2005, 3. Kaplan et al. 2015, 4. Shan et al. 2017, 5. Li et al. 2016a, 6. Wang et al. 2017b, 7. Nakamura et al. 2017a, 8. Nakamura et al. 2017c, 9. Nakamura et al. 2017b, 10. Nakamura et al. 2017d, 11. Nakamura et al. 2018, 12. Xiong et al. 2007, 13. Yang et al. 2013, 14. Chen et al. 2015, 15. Jiang & Mukherjee 2010, 16. Xu et al. 2015.

Appendix C. Summary of ECAS Regimes from the Literature

Based on the literature review given in this chapter, this appendix section briefly summarizes the ECAS regimes that have been successfully tested and used by different authors for the preparation of different transparent or translucent ceramic materials. The listing is in alphabetical order according to the chemical formula and contains the following items, respectively: 1. chemical formula, 2. maximum sintering temperature(s), 3. maximum heating rate(s) and 4. maximum pressure(s). More detailed information can be found in the main text.

$Al_6Si_2O_{13}$ (3 $Al_2O_3 \cdot$ 2 SiO_2): 1250–1600°C, 100°C/min, 15–80 MPa
Al_2O_3: 900–1680°C, 1–1000°C/min, 50–1000 MPa
AlN: 1800°C, 100–550°C/min, 25–30 MPa
AlON: 1350–1700°C, 100°C/min, 30–40 MPa
$BaTiO_3$: 900°C, 100°C/min, 200 MPa
$SrTiO_3$: 950°C, 100°C/min, 200 MPa
$Ba_xSr_{(1-x)}TiO_3$: 1000°C, 100°C/min, 50 MPa
$BaZr_{0.5}Ce_{0.3}Y_{0.2}O_3$: 1350°C, 100°C/min, 50 MPa
$Ca_3(PO_4)_2$: 800–1000°C, 25°C/min, 60 MPa
$Ca_{10}(PO_4)_6(OH)_2$: 700–1200°C, 250200°C/min, 10–500 MPa
CaF_2: 600–1100°C, 100°C/min, 30–100 MPa
CaO: 1600°C, 70°C/min, 80 MPa
$La_{9.33}Si_6O_{26}$: 1200–1500°C, 50°C/min, 100 MPa
$Lu_2Hf_2O_7$: 1450°C, 10°C/min, 100 MPa
Lu_2O_3: 1000–1800°C, 2–200°C/min, 20–100 MPa
$Lu_2Ti_2O_7$: 1000–1500°C, 10°C/min, 100 MPa
$Lu_3Al_5O_{12}$: 1200–1650°C, 10°C/min, 100 MPa
Lu_3NbO_7: 1200–1550°C, 10°C/min, 100 MPa
$MgAl_2O_4$: 900–1600°C, 1–200°C/min, 64–1000 MPa
MgF_2: 700–750°C, 10–100°C/min, 70 MPa
MgO: 650–1500°C, 10–300°C/min, 30–500 MPa

$Pb_{0.92}La_{0.08}(Zr_{0.65}Ti_{0.35})_{0.98}O_3$: 700–900°C, 33°C/min, 29 MPa
$PbZrO_3$-$PbTiO_3$-$Pb(Zn_{0.33}Nb_{0.66})O_3$: 800°C, 33°C/min, 29 MPa
Si_3N_4 and SiAlON: 1850°C, 100°C/min, 30 MPa
$Sr_2Y_8(SiO_4)_6O_2$: 1400–1500°C, 100°C/min, 23 MPa
$Sr_5(PO_4)_3F$: 1050°C, 50°C/min, 100 MPa
$SrHfO_3$: 1700°C, 50°C/min, 100 MPa
Y_2O_3: 850–1600°C, 2–112.5°C/min, 6–300 MPa
Y_2O_3-MgO: 1000–1600°C, 80–200°C/min, 50–80 MPa
$Y_3Al_5O_{12}$: 700–1550°C, 2–200°C/min, 0.5–100 MPa
$ZnAl_2O_4$: 1200–1275°C, 10–100°C/min, 80–100 MPa
ZnO: 900–1100°C, 5–200°C/min, 40–100 MPa
ZnS: 850°C, 5–100°C/min, 80 MPa
ZrO_2: 1050–1350°C, 5–100°C/min, 40–400 MPa

INDEX

D

E

F

G

M

N

O

P

R

S

T

U

V

W

Y

Z